Enjoy the beauty and
diversity of Oman's plants

Annette Patzelt
6. 6. 2011

FIELD GUIDE TO THE

WILD PLANTS OF OMAN

FIELD GUIDE TO THE WILD PLANTS OF OMAN

Helen Pickering and Annette Patzelt

Kew Publishing
Royal Botanic Gardens, Kew

Kew

PLANTS PEOPLE
POSSIBILITIES

All photography by Helen Pickering and Annette Patzelt

First published in 2008 by
Royal Botanic Gardens, Kew
Richmond, Surrey, TW9 3AB, UK

www.kew.org

ISBN 978 1 84246 177 8

British Library Cataloguing in Publication Data
A catalogue record for this book is available from the British Library.

Production editor: Michelle Payne
Typesetting and page layout: Margaret Newman
Design by Media Resources, Royal Botanic Gardens, Kew
Royal Botanic Gardens, Kew

Printed and bound in the UK by Henry Ling Limited

For information or to purchase all Kew titles please visit
www.kewbooks.com or email publishing@kew.org

All proceeds go to support Kew's work in saving the world's plants for life

FSC
Mixed Sources
Product group from well-managed
forests and other controlled sources
www.fsc.org Cert no. SA-COC-001860
© 1996 Forest Stewardship Council

The paper used in this book contains wood from well-managed forests, certified in accordance with the strict environmental, social and economic standards of the Forest Stewardship Council (FSC).

Contents

Acknowledgements

In Oman the authors would like to thank: the staff of the Department of Biology at the Sultan Qaboos University, including Dr Saif Al-Bahry, Dr Mehdi Jaafar, Dr Derek Robinson and Dr Abdul El- Shafie; Sadiqa Ramdan and Jacqueline Henrot of the National Herbarium in the Ministry of National Heritage and Culture for their help with identifications; and Hatim Al-Taie and Rajan Pillai for many helpful suggestions. We acknowledge the role of Andrew Spalton from the Office of the Adviser for Conservation of the Environment, Diwan of Royal Court and the staff in his office. A big thank you goes to David Insall for his encouragement and many useful suggestions throughout the project.

We are grateful to Hossein Akhani from the University of Tehran for expertise in the identification of *Heliotropium*. From the Royal Botanic Garden, Edinburgh we thank Antony Miller and Sabina Knees, who helped with identifications. We are grateful to Fiona Dennis, Annette Jepson, and William Russell for their contributions. A very special thank you goes to Shahina Ghazanfar at Royal Botanic Gardens, Kew to whom we are indebted for her careful reading and comments on the final manuscript and whose constant enthusiasm and help have been invaluable.

Introduction

The Sultanate of Oman is located in the southern part of the Arabian Peninsula and, with the exception of the north western tip, lies within the tropic of Cancer. It has an estimated land area of 309,500 km^2 and 1,700 km of coast. The Arabian Peninsula once formed a land bridge linking Ethiopia and Somalia with south-west Asia. The vegetation therefore contains elements of both regions: the northern Hajar mountains mainly having plants similar to south-west Asian flora, while the central desert and southern Oman (Dhofar) more commonly have flora representative of north and east Africa. There is, however, considerable overlap, with many species occurring in several locations.

The main floristic diversity is found in Dhofar and the northern mountains. The central plains and desert areas have a limited range of plants. Nearly half of the flowering plants are annuals which bloom very briefly after rain. In Dhofar this is mainly during and directly after the monsoon, from August to October. In the northern mountains, where rain is mainly during the winter months, flowering is generally in the spring, from February to April.

SULTANATE OF OMAN

Produce by National Survey Authority
This map is not an authority on international boundaries

The northern Hajar mountains

The rugged Hajar mountains in northern Oman stretch for 700 km from the Musandam Peninsula to Ras al Hadd in the east. The Samail gap, through which the main north–south road runs, divides the range into the eastern and western Hajar. The Jabal Al Akhdar range in the western Hajar reaches altitudes of over 3,000 m, with Jabal Shams (3,009 m) being the highest point. Here annual rainfall is about 350 mm, falling mainly in the winter months. At lower altitudes annual rainfall decreases to an average of 75 mm.

Above 2,000 m the exposed, rocky slopes are dominated by gnarled juniper and olive trees with associated species. Temperate fruit such as plums, peaches and pomegranates are cultivated on the plateau in the Jabal Al Akhdar range. Between 1,000–2,000 m open woodlands have scattered olive, *Sideroxylon* and *Dodonea* trees and shrubs.

Wadis

The mountains of the Arabian Peninsula are deeply dissected by wadis. The Arabic term wadi refers to valleys which act as intermittent rivers, or trace the course of former drainage channels. They generally have rough, fairly flat gravel surfaces which form communication routes between villages. Some contain springs and have perennial water courses and pools, but the majority only have water after rainfall, which may have fallen in distant mountains and can cause flash floods. Even dry wadis generally have water near the surface throughout the year, and therefore have been populated for centuries with small villages where food crops are grown and livestock raised.

The rocky surface of wadis means they are not generally browsed by goats and camels and, as water is often available, many smaller annuals and perennials are found in these areas.

Oasis settlements

Parts of the more fertile wadis and mountains have supported agricultural communities for centuries. These oasis areas are man-made, contain accumulated fine soil, have high, narrow terraces on steep slopes and are irrigated; often using the 'falaj' system, which dates from around 3000 BC. At lower altitudes oasis settlements are characterised by date palms interspersed with cereals, vegetables and fruit. At higher altitudes walnuts, figs, pomegranates and soft fruits are grown.

Coasts

Oman has 1,700 km of coast. The 270 km of the northern coast from Musandam to Muscat and the area south of Ras al Hadd to Dhofar are mainly sandy beaches and salt marshes with a few limestone headlands. The north-east coast, from Muscat to Ras al Hadd, is characterised by steep rocky promontories dropping into the Gulf of Oman. The 800 km of the southern Dhofar coast consist of either wide, sandy dunes or, towards the Yemen border, rocky cliffs descending steeply into the Arabian Sea.

Mangroves (*Avicennia marina*) are dominant in scattered areas along parts of the coasts. Other key species in coastal areas are nearly all perennial, generally succulent, semi-woody dwarf shrubs, or prostrate vines spreading across the beaches. Inland saline areas have very little vegetation.

Gravel desert

The inland deserts of Oman, as well as much of the coastal plain and the northern region of Dhofar, beyond the monsoon area, consist of open gravel desert receiving less than 100 mm of rainfall per annum. These areas consist of a rocky substrate of limestone, sandstone and shale, with a sparse covering of soil.

The vegetation consists mainly of tree scrub, principally *Acacia tortilis* and *Prosopis cineraria*, both of which provide important fodder for livestock. The sparse ground cover is mainly of members of the Chenopodiaceae and Zygophyllaceae families with a few grasses. A small number of fast growing annuals germinate in depressions where the limited rainfall collects. The Jidda, in the central gravel desert, is a clearly defined ecological unit of 40,000 km^2. Here irregular precipitation from fog adds considerable moisture to the ground, allowing for greater plant diversity.

Sand desert

There are two main areas of sand desert: the Eastern (Wahiba) Sands, 2–3 hours drive south of Muscat, and the Rub al Khali (Empty Quarter) that runs along the western border with Saudi Arabia and the United Arab Emirates. These areas are characterised by an extremely dry climate and have low structural and plant diversity. Rainfall is less than 50 mm a year resulting in sparse vegetation; limited to a few hardy species such as *Euphorbia riebeckii*, *Tetraena qatarensis* and grasses with scattered groups of *Prosopis cineraria* along wadi channels. Small groups of Bedu and their camels inhabit these areas.

Road sides and disturbed areas

Plants found along roads and in areas disturbed by construction and settlement are usually pioneer species, and as such are important in the process of soil regeneration. They are usually weedy annuals and commonly found in similar environments throughout the tropics. In Oman, pioneer plants include species of *Launaea, Parthenium*, *Heliotropium*, *Solanum*, *Pluchea* and various grasses.

Dhofar

Dhofar in the south has two contrasting climatic zones: the monsoon region along the coast, and the dry gravel desert further north.

The monsoon area stretches for 240 km east from the Yemen border to Jabal Samhan, its highest point at 2,100 m. In the west the mountains drop steeply into the sea and the monsoon area is only about 3 km wide. In the central area of the Salalah plain it reaches its widest point of 30 km. This area falls within the July/August south-western monsoon belt, when half the annual rainfall of 800–900 mm is recorded. Much of this moisture does not fall as rain but precipitates from thick mists that cover the region, allowing a rich woodland vegetation to emerge. Both the mists and the vegetation end abruptly at the watershed, where the woodland gives way to open shrub which then merges with the desert. The low foothills between the coastal plain and the escarpment were originally wooded but are now dominated by drought-tolerant shrubs. Scattered frankincense trees grow in the intermediate area between the monsoon fogs and the desert.

The gravel desert running up to the Saudi Arabian border is known as the Rub al Khali, or Empty Quarter. Just south of this area, adjacent to the small village of Shisr, there is a UNESCO heritage site: 'Ubar'. Some researchers believed this to be the lost city referred to from Pliny onwards as the southern centre of the incense trade. Others suggested that it was an administrative centre, built much later, to protect a permanent water source.

In the early twentieth century there was intense competition among western explorers to cross the Empty Quarter and to find this 'Lost City'. In 1930 Bertram Thomas was the first westerner to do so. His guides pointed out ancient tracks, said to lead to 'Ubar'. Thomas coined the phrase 'Atlantis of the Sands', providing an evocative challenge to subsequent twentieth-century explorers.

With the help of satellite photography, 'Ubar' was located in 1992 and excavations made. Archaeological evidence shows that this site was occupied from 2000 BC to AD 1200. Pottery and artefacts suggest that it was at its height from 500–300 BC with trade links to Mesopotamia, the Classical World and Persia. This trade was probably based on the export of frankincense and myrrh; both of which grow in this area of Arabia. From AD 650–1200 Chinese porcelain and metalwork indicate a resurgence of trade, possibly based on the export of Arabian horses. Today there is a small, fenced, archaeological site and museum on the edge of Shisr (meaning 'cleft' in Arabic).

Organisation of this guide

This photographic field guide is designed to give a flavour of the overall range and beauty of Oman's wild plants. It aims to cover all the geographical regions. Two hundred and fifty species are illustrated, representing twenty percent of the total flora (circa 1,200 species). Short descriptions are given to aid easy identification.

The book is organised by colour, with separate sections for trees and some conspicuous grasses, ferns and mosses. The colour coding is based on that of the most conspicuous feature, which is usually the flowers but in some cases is the fruit, bracts or the calyx. It was not always easy to decide which colour to put some plants into and readers may differ in their opinions on this.

Within these broad categories plants are ordered alphabetically by family, based on *Flowering Plant Families of the World*. Identifications do not go beyond species level. Synonyms are only included for name changes after the year 2000, except for those published in *Plants of Dhofar*. The main focus is on native plants, however a number of introduced species that have become naturalised and are growing wild are included.

Descriptions of the plants come from two main sources: *The Flora of the Sultanate of Oman, volumes I and II*, and *Plants of Dhofar*, with supplementary information from other Floras and websites as listed in the Bibliography. Data on conservation status is from the *Oman Plant Red Data Book*. We have included only features that can be seen with the naked eye.

Common names are given in English and Arabic. The Arabic names are those published in *An Annotated Catalogue of the Vascular Plants of Oman*. Use of the plants refers to their use in Oman, unless otherwise stated. Many of these may be known only to small groups of individuals or in limited localities and may no longer be current, especially in the case of medicinal practices, which are likely to have been replaced by modern healthcare.

We have tried to avoid technical terms where possible; a short glossary of those that do occur is provided towards the end of the book.

Blue

and mauve flowers

Family Acanthaceae (Acanthus Family)

Name

Barleria aucheriana Nees

Common name

thefied

Description

Low woody shrub; up to 50 cm.

Leaves Opposite, grey/green, oval with entire margins, covered in fine white hair.

Flowers Single, pale mauve with five lobes, borne on stalks in the leaf axils.

Fruit/seed Hairy capsule. Seeds swell when wet.

Flowering period

Most of the year.

Habitat

Common on rocky slopes; altitude up to 1,800 m.

Distribution

Widely distributed from N and tropical Africa to India and Pakistan.

Family Acanthaceae (Acanthus Family)

Name

Blepharis ciliaris (L.) B.L.Burtt

Common names
Eyelash plant, nigeyl

Description
Woody perennial with stiff erect or spreading stems; up to 30 cm.

Leaves Opposite, grey/green, stiff with two or three pairs of spines on the margins.

Flowers Purple, borne among spiny bracts, the three lobed lip has dark purple veins.

Fruit/seed Flattened capsule, hidden among spiny bracts. Seeds hairy becoming mucilaginous when wet.

Flowering period
Most of the year.

Habitat
Common in gravel deserts; altitude up to 800 m.

Distribution
Arabia, Iran and Pakistan.

Uses
Reported to have a number of medicinal uses, including stopping bleeding.

Note
Attracts the yellow parsley butterfly (*Precis hierta*).

Family Acanthaceae (Acanthus Family)

Name

Dyschoriste dalyi A.G.Mill.

Common name
dayfed

Description
Prostrate perennial, often rooting at the nodes; up to 60 cm.

Leaves Opposite, spoon-shaped with wavy margins.

Flowers Mauve or pink with dark purple markings on lower lobes, one or two borne in the leaf axils.

Fruit/seed Pale brown capsule, oblong. Seeds pale brown, covered in white hair.

Flowering period
August–November.

Habitat
Common in Dhofar in the wet woodlands; altitude from 250–1,250 m.

Distribution
Endemic to Dhofar and SE Yemen. Similar species are found in E Africa.

Uses
An important fodder plant for cattle and goats.

Note
Named after Ralph Daly, who initiated many environmental measures in Oman in the late twentieth century.

Family Acanthaceae (Acanthus Family)

Name

Hypoestes forskaolii (Vahl) R.Br.

Description
Untidy erect herb with branched stems covered in short hairs; up to 1 m.

Leaves Opposite, elliptic with entire margins.

Flowers Mauve with darker makings on the throat, the upper lip reflexed. Appearing in clusters at the end of axillary or terminal stems.

Fruit/seed Club shaped capsule. Seeds pale brown, smooth.

Flowering period
All year, if near water.

Habitat
The wet escarpment woodlands in Dhofar; altitude from 350–1,100 m.

Distribution
Throughout Africa and Arabia.

Note
Reputed to be toxic to livestock.

Family Acanthaceae (Acanthus Family)

Name

Justicia diffusa Willd.

Description

Spreading, woody perennial, prostrate or erect, stems often reddish; up to 30 cm.

Leaves Opposite, dark green, may have a red tinge, oval with entire margins.

Flowers Pink/white with darker pink markings on the petals, borne on erect spikes.

Fruit/seed Pale brown capsule with four flattened, brown seeds.

Flowering period

September–January.

Habitat

Dry and open habitats in Dhofar; altitude from 500–1,500 m.

Distribution

Widely distributed in W Africa, Ethiopia, Arabia, and India to Australia.

Note

Toxic to goats. There are five further *Justicia* species in Oman. The genus *Justicia* is named after James Justice, an eighteenth-century Scottish horticulturalist.

Family Acanthaceae (Acanthus Family)

Name

Megalochlamys violacea (Vahl) Vollesen

Description

Erect perennial; up to 1 m.

Leaves Opposite, elliptic, with entire margins.

Flowers Blue, four free petals emerging from hairy bracts.

Fruit/seed A capsule with smooth shiny seeds.

Flowering period

July–November.

Habitat

Wet areas of the Dhofar escarpment; altitude up to 1,300 m.

Distribution

NE and E Africa and tropical Arabia.

Family Acanthaceae (Acanthus Family)

Name

Neuracanthus spinosus Deflers

Description

A sub-shrub with prostrate white stems; 17 cm high, spreading up to 1 m.

Leaves Opposite with wavy margins, young leaves whitish and densely hairy, older leaves grey/green.

Flowers Purple/blue, with a white centre and dark needle-like spines beneath.

Fruit/seed A capsule.

Flowering period

August–November.

Habitat

Rocky areas in Dhofar; altitude up to 1,200 m.

Distribution

Yemen and southern Oman.

Family Apocynaceae: Asclepiadoideae (Oleander or Milkweed Family)

Name

Calotropis procera (Aiton) W.T.Aiton

Common names
Sodom's apple, giant milkweed, ashkhar

Description
Shrub, branched from the base with milky latex; up to 5 m.

Leaves Opposite, pale green, oval with entire margins, covered in soft white hair.

Flowers Five pointed petals, purple towards the tip and pale mauve or white near the centre. In groups on stalks either terminally or in the leaf axils.

Fruit/seed Paired capsules. Seeds with a tuft of long hair at one end.

Flowering period
Most of the year.

Habitat
Common on sandy soils and gravel plains; altitude up to 500 m.

Distribution
Pan-tropical.

Uses
Various traditional medicinal uses are reported, including the treatment of infected wounds and skin problems.

Note
Very toxic, unpalatable to livestock. An indicator of overgrazing. Attracts the plain tiger butterfly (*Danaus chrysippus*).

Family Asteraceae (Daisy or Sunflower Family)

Name

Echinops spinosissimus Freyn

Common names
Globe thistle, kan'a, qan'a

Description
Erect perennial with grooved, bristly stems; up to 2 m.

Leaves Concentrated in a basal rosette, with white midribs, deeply toothed margins and white spines.

Flowers Terminal, large spherical heads, pale blue with spiny bracts pointing downwards, immature flower heads have radiating spines.

Fruit/seed Achenes, with short scales.

Flowering period
Most of the year.

Habitat
Common along roadsides and in the northern mountains; altitude from 100–2,000 m.

Distribution
The Mediterranean, Arabia, Central Asia and tropical Africa.

Family Asteraceae (Daisy or Sunflower Family)

Name

Pluchea arabica (Boiss.) Qaiser & Lack

Description

Dense clump forming perennial, the whole plant is aromatic and covered in hair; up to 70 cm high, spreading to 1 m.

Leaves Alternate, elliptic with an acute tip and irregularly serrated margins.

Flowers Compact, pink, terminal, borne on leafless stalks.

Fruit/seed Achenes, covered in bristles.

Flowering period

All year.

Habitat

Dry areas where the water table is near the surface; altitude up to 750 m.

Distribution

This species is found only in southern Arabia and Somalia. Other *Pluchea* species are found throughout the tropics.

Note

Two other *Pluchea* species occur in Oman. An indicator of the presence of water.

Family Asteraceae (Daisy or Sunflower Family)

Name

Vernonia cinerea (L.) Less.

Common names
Little ironweed, kibb

Description
Erect annual or perennial, stems hairy; up to 50 cm.

Leaves Bright green, oval with slightly wavy margins.

Flowers Purple, borne on erect, terminal sprays.

Fruit/seed Achenes with tufts of white hair.

Flowering period
Most of the year.

Habitat
A weed of cultivation and disturbed areas; altitude up to 650 m.

Distribution
Widespread in the Old World tropics.

Uses
A number of medicinal properties are reported, including treatment for scorpion stings.

Family Brassicaceae (Cabbage or Mustard Family)

Name

Physorhynchus chamaerapistrum (Boiss.) Boiss.

Common names
khafij, kawfig

Description
A tall, woody perennial; up to 1 m.

Leaves Alternate, large lower leaves are fleshy, oval with entire margins, upper leaves are smaller and clasp the stem.

Flowers Four free pink/purple petals with darker veins.

Fruit/seed Round, with conspicuous nerves.

Flowering period
January–April.

Habitat
Sand and gravel plains, along roadsides and open areas in northern Oman; altitude up to 900 m.

Distribution
Arabia, Iran and SW Pakistan.

Uses
Powdered seeds mixed with milk are said to remove warts. Favoured as a fodder by goats.

Note
Attracts swallowtail (*Papilio machaon*) and lime (*Papilio demoleus*) butterflies.

Family Commelinaceae (Spiderwort Family)

Name

Commelina forskaolii Vahl

Common name
Rat's ear

Description
Sprawling, slightly succulent perennial; up to 30 cm.

Leaves Alternate, fresh green, lance shaped with entire margins and parallel veins.

Flowers Bright blue with two free petals above and one below.

Fruit/seed Green capsule with white hairs.

Flowering period
Most of the year.

Habitat
Damp areas in wadis and woodlands in Dhofar; altitude up to 1,000 m.

Distribution
Tropical Africa, Arabia and India.

Note
Named after Jan and Caspar Commelin, seventeenth century Dutch botanists.

Family Convolvulaceae (Morning-glory or Bindweed Family)

Name

Ipomoea pes-caprae (L.) R.Br.

Common names
Horse footprint, dibaiya

Description
Trailing perennial with long stems; spreading up to 5 m.

Leaves Alternate, shallowly bi-lobed with entire margins.

Flowers Large, mauve, trumpet shaped with fused petals.

Fruit/seed Roughly rounded capsule. Seeds covered in brown velvety hair.

Flowering period
Most of the year.

Habitat
Forms mats along sandy beaches. May climb over small shrubs; altitude up to 50 m.

Distribution
Pan-tropical.

Note
Reported to be toxic to camels, which become disorientated. *Stictocardia macalusoi* has a similar flower but the leaves have an indented base and acute tip.

Family Lamiaceae (Mint Family)

Name

Lavandula dhofarensis A.G.Mill.

Common name
Lavender

Description
Erect, perennial with square stems; up to 50 cm.

Leaves Opposite with up to five paired segments, hairy.

Flowers Lilac, arranged on elongated spikes.

Fruit/seed Four oblong nutlets.

Flowering period
All year.

Habitat
Drier, rocky areas of the Dhofar escarpment; altitude up to 1,400 m.

Distribution
This species is endemic to Dhofar and SE Yemen. Other *Lavandula* species are widely distributed from the Mediterranean to India.

Uses
A deodorant and perfume.

Note
There are five species of *Lavandula* in Oman, three of which are endemic to Dhofar. The name is from the Latin *lavo* meaning 'to wash', as it was used to perfume baths.

Family Lamiaceae (Mint Family)

Name

Ocimum dhofarense (Sebald) A.J.Paton

Synonym
Becium dhofarense Sebald

Description
Shrub, stems round or square; up to 2 m.

Leaves Opposite, oval, with an acute tip and serrated margins.

Flowers Mauve, in whorls on an erect spike, with long protruding stamens.

Fruit/seed Pale brown nutlets.

Flowering period
July–November.

Habitat
Common on the Dhofar escarpment, often under *Anogeissus dhofarica* trees; altitude from 200–900 m.

Distribution
Endemic to Dhofar and SE Yemen.

Uses
Has a number of uses related to skin conditions.

Family Lamiaceae (Mint Family)

Name

Ocimum forskaolii Benth.

Common names
Basil, rehan, sawma

Description
Clump forming perennial with square stems and short hairs, pleasantly aromatic; up to 40 cm.

Leaves Opposite, oval with very finely serrated margins and a pointed tip.

Flowers Pale mauve to purple, in whorls of six flowers at intervals along the erect spike.

Fruit/seed 3–4 oblong nutlets.

Flowering period
Most of the year.

Habitat
Open areas, rocky slopes, along roads and in plantations; altitude from 100–650 m.

Distribution
Arabia and E Africa.

Uses
Various medicinal properties such as alleviating the symptoms of colds are reported, also used as a perfume.

Family Lamiaceae (Mint Family)

Name

Plectranthus barbatus Andrews

Description
Erect, shrubby perennial, whole plant aromatic; up to 60 cm.

Leaves Opposite, oval with slightly serrated margins.

Flowers Purple with an upper and lower lip, borne along the length of an erect spike.

Fruit/seed 3–4 shiny black nutlets.

Flowering period
August–November.

Habitat
Forms clumps among rocks on the grasslands of the wet escarpment in Dhofar; altitude from 150–900 m.

Distribution
Africa to India, often cultivated as an ornamental.

Uses
Smells similar to Vicks and used as a deodorant. Good fodder with high water content.

Note
P. cylindraceus is also widely distributed in Dhofar; it is smaller with a paler, dusty mauve flower.

Family Lamiaceae (Mint Family)

Name

Salvia aegyptiaca L.

Description

An erect annual, aromatic with branching stems; up to 30 cm.

Leaves Opposite, mainly on the lower part of stems, elliptic with wavy margins, hairy, veins depressed.

Flowers White with purple markings, in whorls of 2–6 flowers at intervals along the upper stems.

Fruit/seed Four nutlets with black seeds.

Flowering period

February–May.

Habitat

Common in wadis and rocky slopes; altitude up to 2,000 m.

Distribution

N Africa, Mediterranean, Middle East, Arabia, Pakistan and India.

Uses

A number of unrelated medicinal uses are reported, mainly from UAE.

Note

There are several other *Salvia* species in Oman, two of which are endemic to Dhofar.

Family Lamiaceae (Mint Family)

Name

Teucrium nummularifolium Baker

Description

Erect or decumbent perennial; up to 30 cm.

Leaves Opposite, grey/green, small, broadly oval with serrated margins, hairy.

Flowers Purple with a large lower lip, inflorescence forms a terminal spike, covered in hair giving a woolly appearance.

Fruit/seed Four tiny nutlets with brown seeds.

Flowering period

Most of the year.

Habitat

Open, dry, stony ground in Dhofar; altitude up to 1,100 m.

Distribution

Endemic to Dhofar and SE Yemen.

Family Polygalaceae (Milkwort Family)

Name

Polygala mascatensis Boiss.

Description

Compact perennial herb, woody at the base; 30–40 cm high.

Leaves Alternate, grey/green, small with entire margins, covered in soft white hair.

Flowers Purple with two wing-like sepals on either side. The central of the three petals has a tuft of purple hair on the end.

Fruit/seed Dark brown, compressed capsule with black, hairy seeds.

Flowering period

Most of the year.

Habitat

Gravel and rocky areas; altitude up to 1,900 m.

Distribution

Endemic to Oman.

Note

There are several species of *Polygala* in Oman, all with the characteristic flower shape. The name *Polygala* comes from the Greek meaning 'lots of milk', suggesting that livestock feeding on these species produce plenty of milk.

Family Primulaceae (Primula Family)

Name

Anagallis arvensis L.

Common names
Blue pimpernel, ain al qat, zraig al ain, al auaina, uwaynah

Description
Annual with branching slender stems, sometimes spreading or climbing; up to 30 cm.

Leaves Opposite, oval with entire margins and three parallel veins.

Flowers Single, dark blue with five petals and a red centre.

Fruit/seed Round capsule bending down on a thin stalk.

Flowering period
Most of the year.

Habitat
Damp areas in plantations, wadis and mountains; altitude up to 2,000 m.

Distribution
Widely naturalised in moist tropical and temperate regions.

Note
Anagallis is derived from the Greek meaning 'delightful'.

Family Scrophulariaceae (Foxglove or Figwort Family)

Name

Orobanche ramosa L.

Common name
Branched broomrape

Description
An erect annual, parasitic, stems covered in dense woolly hair; up to 20 cm.

Leaves Scale-like, unobtrusive.

Flowers Bright blue with white centres, trumpet shaped, occurring along the length of the stem.

Fruit/seed A capsule in the persistent corolla. Seeds tiny, up to 40,000 per plant.

Flowering period
January–April.

Habitat
Damp, shady areas and around cultivation; altitude up to 2,600 m.

Distribution
N and E Africa, C and S Europe, Middle East and W Asia, introduced into other tropical countries.

Note
From the Greek *Orobos* meaning 'to strangle'. Parasitizes a wide range of grain and vegetable crops, resulting in a yield reduction of over 25 percent.

Family Solanaceae (Potato Family)

Name

Hyoscyamus insanus Stocks

Description

Branched perennial; up to 60 cm.

Leaves Alternate, large fleshy, concentrated at the base of the plant, with broadly serrated margins and pale veins.

Flowers Large, white streaked with purple, borne on erect spikes.

Fruit/seed An oval capsule.

Flowering period

February–May.

Habitat

Along roadsides and coastal areas; altitude up to 1,500 m.

Distribution

Southern Arabia and east to Pakistan.

Uses

Has a long history of being used as a sedative.

Note

There are two further species of *Hyoscyamus* in Oman, both are endemic.

Family Solanaceae (Potato Family)

Name

Solanum incanum L.

Common names
Palestine nightshade, shirinjiban

Description
Small shrub densely covered in short woolly hair. Stems yellow/green with both curved and straight prickles; up to 1 m.

Leaves Alternate, grey/green, broadly oval with wavy margins, often with spines on the mid-rib.

Flowers Mauve, solitary with yellow stamens.

Fruit/seed Large berry turning from green to yellow as it matures.

Flowering period
Most of the year.

Habitat
Gravel plains, rocky slopes, waste and disturbed areas; altitude up to 2,000 m.

Distribution
Egypt to Ethiopia and east to Pakistan and India.

Uses
Fruit used in tanning hides and medicinally for toothache and bruises.

Note
Toxic, can cause death in animals. Indicator of overgrazing. Attracts the pearl charaxes butterfly (*Stonehamia varans*).

Red

and pink flowers

Family Acanthaceae (Acanthus Family)

Name

Ruttya fruticosa Lindau

Description
Straggling shrub; up to 3 m.

Leaves Opposite, oval with entire margins.

Flowers Bright red/orange, two-lipped with black wrinkled, shiny throats, in groups of up to ten.

Fruit/seed Club shaped with a long stalk. Seeds released explosively when capsule is moistened.

Flowering period
September–November.

Habitat
Seaward facing slopes of the escarpment mountains in Dhofar; altitude from 180–900 m.

Distribution
Tropical Africa to southern Arabia.

Uses
An important forage plant.

Note
Named after Dr John Rutty, an Irish physician (1698–1775).

Family Amaranthaceae (Amaranthus, Celosias or Cockscomb Family)

Name

Achyranthes aspera L.

Common name
harsha

Description
An erect herb, the whole plant may be tinged red; up to 80 cm.

Leaves Opposite, oval, with entire margins, surface either smooth or hairy.

Flowers Reddish bracteoles surround small green flowers, on erect spikes, bristly at the top, drooping lower down.

Fruit/seed Ovoid capsule, when mature prickly bracts attach to passing animals.

Flowering period
Most of the year.

Habitat
Common weed of cultivation, also on rocky ground and in wadis; altitude up to 2,500 m.

Distribution
Native to the Old World tropics, now pan-tropical.

Family Amaranthaceae (Amaranthus, Celosias or Cockscomb Family)

Name

Digera muricata (L.) Mart.

Description
Erect clump-forming annual, may have fine white hairs on stems; up to 70 cm.

Leaves Opposite, linear, with entire margins.

Flowers Pink borne on spikes arising from the leaf axils.

Fruit/seed Round nutlet.

Flowering period
Most of the year.

Habitat
Weed in damp, disturbed areas at the edges of fields or roadsides; altitude up to 1,200 m.

Distribution
Native to the Old World tropics.

Family Apiaceae (Carrot or Parsley Family)

Name

Pycnocycla aucheriana Decne. ex Boiss.

Description An erect shrub with stems ending in spines; up to 60 cm.

Leaves Grey/green, stiff, deeply dissected with 3–5 spiny segments and a central groove at the base of the plant.

Flowers Pale pink, in a compact flower head with 5–8 long spines below.

Fruit/seed Cylindrical.

Flowering period
February–May.

Habitat
Forms large clumps in the foothills of the northern mountains; altitude up to 1,000 m.

Distribution
Northern Oman, Iran and Pakistan.

Note
This plant is becoming rare.

Family Apocynaceae: Asclepiadoideae (Oleander or Milkweed Family)

Name

Adenium obesum Roem. & Schult.

Common name
Desert rose

Description
Small tree or shrub with a large swollen base, bark pale grey, smooth with clear or white sap; up to 1.5 m.

Leaves Alternate, often in terminal clusters, green above and either pale green or reddish below, oval to round with entire margins.

Flowers Bright pink, funnel shaped with five petals, borne at the end of branches.

Fruit/seed Two red cigar shaped follicles fused at the base, seeds have tufts of hair on both ends.

Flowering period
Most of the year, mainly May–June.

Habitat
Dry rocky areas, very common in Dhofar; altitude up to 1,700 m.

Distribution
Native to E Africa, widespread in tropical Africa and southern Arabia.

Uses
Sap used to treat swelling and joint pain. In Africa the sap is used as an arrow poison.

Note
All parts are toxic to animals.

Family Apocynaceae: Asclepiadoideae (Oleander or Milkweed Family)

Name

Desmidorchis arabica (N.E.Br.) Meve & Liede

Synonym

Caralluma arabica N.E.Br.

Common name

khansur

Description

Erect, spreading, succulent perennial with square grey stems with wavy edges; up to 30 cm.

Leaves Absent.

Flowers Tight cluster of dark red flowers, each with a white spot in the centre, forming a ball at the end of the stems.

Fruit/seed Paired, erect, cylindrical with dark brown seeds and a tuft of long hair.

Flowering period

February–May.

Habitat

Rocky areas of the northern mountains; altitude up to 2,200 m.

Distribution

Endemic to northern Oman, UAE and Yemen.

Uses

Reported as a treatment for liver complaints, high blood pressure and diabetes.

Note

Attracts the plain tiger butterfly *(Danaus chrysippus)*. The carrion-like smell attracts flies.

Family Apocynaceae: Asclepiadoideae (Oleander or Milkweed Family)

Name

Desmidorchis penicillata (Deflers) Plowes

Synonym

Caralluma penicillata (Defl.) N.E.Br.

Common name

dij

Description

Erect, succulent with dense pale green/grey branches; up to 40 cm.

Leaves Absent.

Flowers Star shaped, green, covered in purple spots with tufts of purple hair on the tips.

Fruit/seed Cylindrical, paired, erect with dark brown seeds and a tuft of long hair.

Flowering period

February–May.

Habitat

In Oman only found in the Jabal Al Akhdar region, associated with *Juniperus excelsa* and *Olea europaea*; altitude from 1,000–1,900 m.

Distribution

NE Africa and Arabia.

Note

Has a restricted distribution and is at risk of becoming extinct.

Family Apocynaceae: Asclepiadoideae (Oleander or Milkweed Family)

Name

Nerium oleander L.

Common name
Oleander, rosebay, haban

Description
Erect multi-branched shrub, with white latex; up to 4 m.

Leaves Whorled, lance-shaped, with entire margins and prominent midribs.

Flowers Pink, funnel shaped with five petals, borne in clusters at the end of branches.

Fruit/seed Narrow cylindrical pods hang down and split to reveal many seeds with brown tufts of hair.

Flowering period
All year.

Habitat
Forms dense clumps in wadi bottoms with permanent water; altitude up to 1500 m.

Distribution
Native to the Mediterranean, a common ornamental in tropical and semi-tropical regions.

Note
Very poisonous, contains toxic cardiac glycosides.

Family Apocynaceae: Asclepiadoideae (Oleander or Milkweed Family)

Name

Periploca aphylla Decne.

Common name
handabub

Description
Erect almost leafless shrub that produces white latex when cut; up to 3 m.

Leaves Opposite, tiny, soon falling.

Flowers Deep red/purple, borne in axillary clusters with five triangular petals curved backwards and fringed with white hair.

Fruit/seed Often paired. Seeds flat, reddish with a tuft of hair at the top.

Flowering period
November–April.

Habitat
Hillsides and wadi banks; altitude up to 2,000 m.

Distribution
Egypt, Arabia, Afghanistan, Pakistan and India.

Uses
Used in Dhofar to clear pools of leeches which are attracted to the plant. Buds are edible. The roots are sometimes used for snakebite.

Note
Attracts the desert white butterfly *(Pontia glauconome)*.

Family Aristolochiaceae (Birthwort Family)

Name

Aristolochia bracteolata Lam.

Common name
ghaghae

Description
A prostrate or climbing perennial with an unpleasant smell; up to 20 cm.

Leaves Alternate, grey/green, heart shaped with wavy margins.

Flowers Solitary, pipe shaped with a yellow tube and dark red hairy lip.

Fruit/seed Oblong capsule with 12 ribs and many triangular seeds.

Flowering period
Most of the year.

Habitat
A weed in shady and damp places, near permanent water courses and on the edge of cultivated areas; altitude up to 1,700 m.

Distribution
Tropical Africa, Arabia, Sri Lanka, India and Pakistan.

Uses
Reported as a treatment for snake and scorpion bites in Oman and as an anti-helminth in Pakistan.

Family Asphodelaceae (Aloe Family)

Name

Aloe dhufarensis Lavranos

Description

A succulent, perennial herb; up to 1 m.

Leaves Spirally arranged as a rosette at the base of the stem, grey, large, succulent, slightly incurved with a sharp tip. Margins generally entire or with a few small teeth near the base.

Flowers On an erect stalk, up to 1 m high, carrying many upright red buds which become pendulous red/yellow when open.

Fruit/seed Erect, with three valves, hard when mature. Seeds brown and winged.

Flowering period

Most of the year.

Habitat

Common in drier areas and on north facing slopes of the escarpment in Dhofar; altitude up to 1,400 m.

Distribution

Endemic to Dhofar.

Uses

Considered to be one of the most important medicinal plants in Dhofar, particularly the dried juice.

Note

Has a restricted distribution and therefore at risk of becoming extinct.

Family Asteraceae (Daisy or Sunflower Family)

Name

Dicoma schimperi (DC.) Baill. ex O.Hoffm.

Description

Woody based annual with grooved stems; up to 70 cm.

Leaves Alternate, grey/green, oval with a pointed tip and rough edges.

Flowers Mass of cream/pink flower heads at the end of stems.

Fruit/seed Achene with white hair.

Flowering period

February–May.

Habitat

Mountain slopes and wadis; altitude from 100 to 750 m.

Distribution

Africa and Arabia.

Uses

A yellow dye is made from the crushed leaves in UAE.

Family Asteraceae (Daisy or Sunflower Family)

Name

Kleinia saginata P.Halliday

Description

Succulent perennial with a thick green cylindrical stem; up to 40 cm.

Leaves Often tinged red, small, soon falling leaving white scars.

Flowers A mass of bright red flowers in solitary heads on long stalks at the end of the stem.

Fruit/seed Oblong achenes with white hair.

Flowering period

June–November.

Habitat

Vertical cliffs and ledges on rocky slopes just beyond the monsoon area of Dhofar; altitude up to 1,250 m.

Distribution

Endemic to Dhofar and SE Yemen.

Uses

Various medicinal uses are reported, most commonly as a haemostatic.

Note

The epithet '*saginata*' comes from the Latin, 'to fatten', and refers to the stems.

Family Asteraceae (Daisy or Sunflower Family)

Name

Phagnalon viridifolium Decne. ex Boiss.

Description

Small woody perennial with conspicuous white, hairy, spreading stems; up to 30 cm.

Leaves Alternate, linear with wavy margins.

Flowers Reddish/brown compact heads on long stalks, surrounded by brown bracts, never fully open.

Fruit/seed A cylindrical achene.

Flowering period

February–April.

Habitat

Stony hillsides in northern Oman; altitude from 1,000 to 2,900 m.

Distribution

Endemic to Arabia. Other *Phagnalon* species are found in the Mediterranean region.

Family Asteraceae (Daisy or Sunflower Family)

Name

Pluchea dioscoridis (L.) DC.

Common name
Ploughman's spikenard

Description
An erect perennial, usually hairy, may lean over; up to 3 m.

Leaves Alternate, rough, oblong, with irregularly serrated margins.

Flowers Flower heads cluster at the end of leafless stalks, have a pink tinge, aromatic.

Fruit/seed Hairy achene with bristles on top.

Flowering period
November–April.

Habitat
Disturbed sandy areas and as a weed of cultivation; altitude up to 150 m.

Distribution
Pan-tropical.

Note
Named after the first-century Greek herbalist Dioscorides. Attracts the plain tiger butterfly (*Danaus chrysippus*).

Family Balsaminaceae (Balsam, Jewel-weed or Touch-me-not Family)

Name

Impatiens balsamina L.

Description

Erect annual herb with simple fleshy stems; up to 1.5 m.

Leaves Opposite, elliptic with serrated margins, spirally arranged.

Flowers Pink/mauve, may be in clusters of up to three, five petals with the two lower ones fused into a lip.

Fruit Ovoid, covered in fine hair.

Flowering period

August–November.

Habitat

Very common on the Dhofar escarpment; altitude from 100 to 850 m.

Distribution

Native to India and SE Asia, widely naturalised in Africa.

Uses

Reported as a treatment for eye infections and used as soap.

Family Caryophyllaceae (Pink or Carnation Family)

Name

Silene conoidea L.

Description
An erect annual with hairy stems; up to 30 cm.

Leaves Opposite, lance-shaped with entire margins.

Flowers Five free pink petals.

Fruit/seed Reddish, flask-shaped capsule.

Flowering period
February–April.

Habitat
Between rocks and on rocky slopes, a weed of cultivation in northern Oman; altitude up to 1,500 m.

Distribution
Cosmopolitan.

Note
There are four other *Silene* species in Oman.

Family Chenopodiaceae (Goose Foot, Beetroot or Spinach Family)

Name

Halopeplis perfoliata (Forssk.) Bunge ex Asch.

Common names
String of beads, gharaz (meaning pearls)

Description
Branched, somewhat woody perennial, older stems leafless; up to 50 cm.

Leaves Succulent, often bright red, surround the stem giving the appearance of swollen nodes.

Flowers Minute, fleshy, in clusters between the stem joints.

Fruit/seed Nutlet with tiny seeds.

Flowering period
October–February.

Habitat
Along coastal and inland salt flats; altitude up to 100 m.

Distribution
Palestine, Arabia, Sinai and east to Pakistan.

Note
Tolerant of high salinity.

Family Chenopodiaceae (Goose Foot, Beetroot or Spinach Family)

Name

Salsola drummondii Ulbr.

Description

Erect, woody shrub with greyish bark, older branches become gnarled; up to 70 cm.

Leaves Opposite, stubby, succulent with entire margins.

Flowers Inconspicuous, interspersed with leaves, surrounded by succulent leaf-like pink, papery bracts.

Fruit/seed Round with yellow or pink papery wings.

Flowering period

March–April.

Habitat

Coastal and inland saline sandy areas in northern and central Oman; altitude up to 150 m.

Distribution

Arabia, Iran, Pakistan and India.

Family Chenopodiaceae (Goose Foot, Beetroot or Spinach Family)

Name

Suaeda vermiculata Forssk. ex J.F.Gmel.

Description
Spreading, much branched succulent shrub, woody at the base, stems may be reddish; up to 1.5 m.

Leaves Alternate, grey/green, fleshy, cylindrical, succulent.

Flowers Clusters of green flowers with red stamens, borne in the leaf axils.

Fruit/seed Seeds are small, black and shiny.

Flowering period
Most of the year.

Habitat
Common in sandy and saline areas, along the coast and the edges of cultivation; altitude up to 1,200 m.

Distribution
Africa to SE Asia.

Uses
Smoke from burning leaves is said to relieve symptoms of asthma.

Family Dracaenaceae (Dragon-Tree Family)

Name

Sansevieria ehrenbergii Schweinf. ex Baker

Description
Succulent perennial with no stem; up to 1 m.

Leaves Long, narrow, arising from a creeping rhizome, mature leaves may reach 1m in length, are thick and rigid, more or less circular in section with a long groove running the length of the leaf to the acute tip.

Flowers Dark red/brown with white stamens, grouped on a tall spike.

Fruit/seed A hard berry with few seeds.

Flowering period
May–June.

Habitat
Common on low dry hills and coastal hills in Dhofar; altitude from 50–600 m.

Distribution
Africa and southern Arabia.

Uses
Grown as an ornamental. Fibres are used to make rope.

Family Euphorbiaceae (Spurge Family)

Name

Euphorbia riebeckii Pax

Description

Dense low growing, spiny mound with zig zagging reddish stems and milky sap; up to 80 cm.

Leaves Opposite, grey, small with entire margins.

Flowers Small, red, borne at the end of the branches.

Fruit/seed A round three-lobed capsule.

Flowering period

February–April.

Habitat

Coastal and inland sandy areas; altitude up to 350 m.

Distribution

Endemic to Oman and Yemen.

Family Geraniaceae (Geranium or Pelargonium Family)

Name

Geranium mascatense Boiss.

Common names
Geranium, shuwayb al hamam

Description
Annual herb with erect or spreading stems; up to 40 cm.

Leaves Opposite, deeply segmented into three or five lobes, margins serrated, covered in soft hair.

Flowers Pink with a dark purple centre.

Fruit/seed Five one seeded sections, deeply wrinkled.

Flowering period
After rain in the northern mountains and Dhofar.

Habitat
Shady areas, often on the edges of irrigated plantations; altitude up to 2,200 m.

Distribution
E Africa, Arabia, Iran and Pakistan.

Family Iridaceae (Iris Family)

Name

Gladiolus italicus Mill.

Common names
sa'ad, maash

Description
A perennial herb; up to 1 m.

Leaves Alternate, lance-shaped, long with entire margins.

Flowers Up to 12 pink flowers borne on one side of the stem.

Fruit/seed A round capsule between two bracts. Seeds round.

Flowering period
February–April.

Habitat
On the edges of cultivation in northern Oman; altitude from 1,330 to 1,500 m.

Distribution
Mediterranean, Oman, Saudi Arabia, UAE, introduced into many tropical and subtropical regions.

Family Lamiaceae (Mint Family)

Name

Teucrium stocksianum Boiss.

Common name
ja'ada

Description
Small, dense, aromatic perennial; about 25 cm high.

Leaves Opposite, grey/green, small, cylindrical, covered in soft white hair.

Flowers A terminal cluster of pale yellow/pink flowers with dark pink veins, lower lip larger than the two upper ones.

Fruit/seed Four tiny nutlets.

Flowering period
Most of the year.

Habitat
Rocky hillsides in northern Oman; altitude up to 1,900 m.

Distribution
Oman, UAE, Iran and Pakistan.

Uses
Some medicinal uses reported, mainly for abdominal problems.

Family Leguminosae (Pea, Bean or Mimosa Family)

Name

Indigofera spp.

Description

A large genus of over 700 species. Range from small prostrate annuals to trees up to 3 m.

Leaves Usually compound, may be simple. Broadly oval with entire margins. Often grey/green.

Flowers Red or pink, pea-like, borne in the leaf axils.

Fruit/seed Pods, often flat, may be curved or straight.

Flowering period

After rain.

Habitat

Sunny, open, often rocky areas; altitude up to 2,000 m.

Distribution

Widely distributed in tropical and sub-tropical regions. Used as dyes. There are 15 species of *Indigofera* in Oman.

Name

Indigofera arabica Jaub. & Spach

A small prostrate perennial with red flowers and flattened pods.

Name

Indigofera caerulea Roxb.

A hairy shrub up to one metre in height with bluish leaves, red flowers and curved pods. Formerly a prime source of the dye indigo.

Family Leguminosae (Pea, Bean or Mimosa Family)

Name

Indigofera volkensii Taub.

Rough leaves and relatively large flowers. Pods are straight.

Name

Indigofera intricata Boiss.

A small cushion-forming shrub; up to 70 cm. Branches end in spines, whole plant covered in silky white hair. Leaflets and flowers are tiny.

Family Leguminosae (Pea, Bean or Mimosa Family)

Name

Lotus garcinii DC.

Description
Dense clump forming perennial with hairy stems; up to 80 cm.

Leaves Alternate, grey/green, small, compound, each having 3–5 leaflets, widely spaced on the stems.

Flowers Solitary pink flowers have darker veins, inconspicuous, being almost hidden in the hairy calyx.

Fruit/seed An oblong pod, green/brown with six seeds.

Flowering period
December–April.

Habitat
Sandy and often saline areas; altitude sea level.

Distribution
Somalia, Egypt, southern Arabia, Iran and Pakistan.

Family Leguminosae (Pea, Bean or Mimosa Family)

Name

Taverniera cuneifolia (Roth) Arn.

Common name
asmat

Description
Woody shrub with green stems branched from the base; up to 1 m.

Leaves Opposite, may have three leaflets, very sparse, small with entire margins and a rounded tip.

Flowers Pink with dark veins in clusters in the leaf axils.

Fruit/seed Hairy pod with up to three joints. Seeds dull brown or greenish.

Flowering period
January–May.

Habitat
Sandy, gravel and rocky areas in northern Oman; altitude up to 800 m.

Distribution
NE Africa, UAE, Iran, Pakistan and India.

Note
Named after Jean Tavernier, a French traveller in the seventeenth century.

Family Leguminosae (Pea, Bean or Mimosa Family)

Name

Tephrosia nubica (Boiss.) Baker

Common name
aytiman

Description
Dense woody perennial with erect branching stems, all parts covered in fine white hair; up to 1.25 m.

Leaves Alternate, greyish, compound with 4–8 pairs and a terminal leaflet, widely spaced on a central stalk, with entire margins.

Flowers Pink with a hairy upper petal, borne on stems arising from the leaf axils or terminally.

Fruit/seed Oval pods covered in long yellowish hair. Seeds kidney-shaped.

Flowering period
December–May.

Habitat
Common in gravel plains and wadis in northern Oman; altitude up to 400 m.

Distribution
Tropical Africa and Arabia.

Family Loranthaceae (Mistletoe Family)

Name

Plicosepalus acaciae (Zucc.) Wiens & Polh.

Synonym
Loranthus acaciae Zucc.

Description
A parasitic shrub, dependent on the host plant for water and nutrients.

Leaves Alternate in clusters, oblong, thick and leathery with entire margins.

Flowers Bright red, with reflexed lobes, clustered in the leaf axils.

Fruit/seed A sticky red berry with a single seed.

Flowering period
September–October.

Habitat
Mainly in drier areas of Dhofar, often parasitic on *Acacia* and *Commiphora* trees; altitude from 500–900 m.

Distribution
NE Africa and Arabia.

Uses
Tanning leather.

Note
Attracts the Arabian sapphire (*Epamora glaucus*) butterfly.

Family Lythraceae (Loosetrife, Pomegranate, Henna, Water Chestnut or Crepe Myrtle Family)

Name

Ammannia baccifera L.

Description
Erect, very branched annual with square stems; up to 40 cm.

Leaves Opposite, oblong with entire margins, reflexed.

Flowers Red, clustered in the leaf axils along the length of the stem.

Fruit/seed A red capsule with brownish seeds.

Flowering period
January–April.

Habitat
Mainly in shady, wet or marshy places; altitude up to 1,000 m.

Distribution
Pan-tropical weed.

Note
Salt tolerant.

Family Malvaceae (Mallow, Cotton or Hollyhock Family)

Name

Gossypium stocksii Mast.

Description

Prostrate woody herb with slightly hairy stems; spreading up to 2 m.

Leaves Alternate, palmately lobed with entire margins, dotted with black glands.

Flowers Solitary, pink with five spirally arranged fused petals and a red centre, borne in the leaf axils.

Fruit/seed Round capsule, dotted with black glands, enclosed in enlarged epicalyx lobes.

Flowering period

Most of the year.

Habitat

Common on dry coastal plains and rocky slopes in Dhofar; altitude up to 150 m.

Distribution

This species is only found in Dhofar, Somalia and Pakistan. The genus *Gossypium* is native to dry areas of the tropics.

Uses

Cotton is produced from *Gossypium* species in other parts of the tropics.

Note

G. stocksii is named after Dr J. E. Stocks, who made important plant collections in Pakistan in the nineteenth century.

Family Nyctaginaceae (Bougainvillea or Four o'clock Family)

Name

Boerhavia elegans Choisy

Description
Woody based perennial with red, slender, erect stems; up to 60 cm.

Leaves Opposite, grey with reddish tinge, oblong with entire margins, concentrated at the base of the plant.

Flowers Tiny, red, at the end of branched, erect red stems forming a mist-like spray of colour.

Fruit/seed Club-like with five ribs.

Flowering period
September–May.

Habitat
Common along roadsides and on gravel plains; altitude up to 800 m.

Distribution
Dry areas from NE Africa to SW Asia.

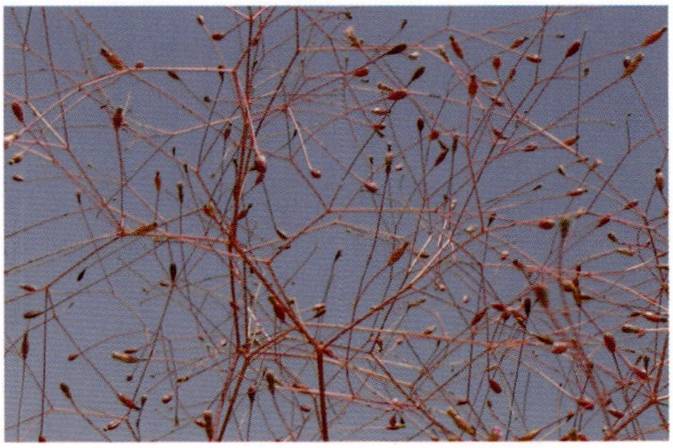

Family Nyctaginaceae (Bougainvillea or Four o'clock Family)

Name

Commicarpus boissieri (Heimerl) Cufod.

Description
Woody perennial with long slim stems; up to 1 m.

Leaves Opposite, dull green, oval with entire margins, young leaves may be hairy with an indented tip.

Flowers Dark pink, funnel shaped with five fused petals, arranged in groups of 3–6 on long stems.

Fruit/seed Cylindrical, covered in prominent glands.

Flowering period
Most of the year.

Habitat
Common in rocky open areas; altitude up to 1,000 m.

Distribution
India, Pakistan and Yemen.

Uses
The roots are edible. Plant heavily grazed by livestock.

Family Plumbaginaceae (Leadwort, Sea-lavender or Thrift Family)

Name

Dyerophytum indicum (Gibs. ex Wight) Kuntze

Common name
malihlah

Description
Erect shrub covered in whitish powder; up to 2 m.

Leaves Alternate with the base clasping the stem, pale green/bluish, leathery with entire margins.

Flowers Red striped funnels with five free, white petals with red stripes in the centre, arranged in spikes at the end of the branches.

Fruit/seed Winged single seed.

Flowering period
Most of the year.

Habitat
Widely distributed in wadis and on rocky slopes; altitude from 150–1,800 m.

Distribution
Southern Arabia and W India.

Uses
Whole plant is covered in salty powder and is sometimes used in cooking as a substitute for salt.

Family Plumbaginaceae (Leadwort, Sea-lavender or Thrift Family)

Name

Limonium axillare (Forssk.) Kuntze

Common name

Sea lavender

Description

Clump forming perennial shrub with erect stems; up to 60 cm.

Leaves Grey/green, spoon shaped with entire margins, often in basal rosettes, covered with salt excreted from glands on the leaves.

Flowers Pale pink to purple sprays rising up to 25 cm above the base, each flower has five petals fused at the base.

Fruit/seed Ovoid, one seeded, dispersed with the calyx.

Flowering period

Most of the year.

Habitat

Common in coastal areas in Dhofar, never far inland; altitude sea level.

Distribution

NE Africa, Arabia, and east to Pakistan.

Uses

Roots can be crushed and used as a dye and to strengthen fishing nets and ropes.

Note

Three other *Limonium* species are found in Oman.

Family Polygonaceae (Dock, Buckwheat or Rhubarb Family)

Name

Polygonum glabrum Willd.

Synonym
Persicaria glabra (Willd.) M.Gomez

Common name
Dense flowered knotweed

Description
Aquatic perennial in dense stands, stems erect, rarely branched; up to 1 m.

Leaves Alternate, bright green, lance-shaped with entire margins.

Flowers Pink clusters, on long terminal spikes.

Fruit/seed Ovoid, dark brown/black, shiny nut.

Flowering period
Most of the year.

Habitat
Common in freshwater stands in Dhofar; altitude from 100–750 m.

Distribution
Pan-tropical.

Family Polygonaceae (Dock, Buckwheat or Rhubarb Family)

Name

Pteropyrum scoparium Jaub. & Spach

Common name
sidaf

Description
Erect, much branched perennial with grey stems; up to 150 cm.

Leaves Succulent, linear, in clusters along the stem.

Flowers Alternate, small, white with red stamens, borne in the leaf axils.

Fruit/seed Nutlet surrounded by conspicuous reflexed papery wings, turning from cream to bright red.

Flowering period
January–April.

Habitat
Wadis and gravel areas in northern Oman, often in large stands; altitude up to 600 m.

Distribution
Regional endemic to UAE and Oman.

Family	Polygonaceae (Dock, Buckwheat or Rhubarb Family)

Name

Rumex vesicarius L.

Common name
Sorrel, bladderdock, hummad

Description
Slightly succulent annual with ribbed stems that may be reddish; up to 30 cm.

Leaves Arrow shaped, often red tinged, on a long stalk with entire margins.

Flowers White, inconspicuous.

Fruit/seed Surrounded by bright red wings with red veins when mature.

Flowering period
Most of the year, if near water.

Habitat
Common along roadsides, sandy and gravel hillsides; altitude up to 1,800 m.

Distribution
N Africa, Arabia and SW Asia.

Note
Leaves contain oxalic acid giving them a lemony taste.

Family Rubiaceae (Coffee or Madder Family)

Name

Pseudogaillonia hymenostephana (Jaub. & Spach) Lincz.

Common name
Popcorn plant

Description
Cushion-forming woody perennial, stems grey; up to 40 cm.

Leaves Opposite, linear with entire margins.

Flowers Clusters of pink flower heads cover the bush. Tiny white flowers with five petals rise above the broad pink calyx.

Fruit/seed A capsule with tiny seeds.

Flowering period
February–April.

Habitat
Rock crevices and mountain slopes of the northern mountains; altitude up to 2,000 m.

Distribution
Arabia and SW Asia.

Family Sapindaceae (Litchi, Maple or Horse chestnut Family)

Name

Dodonaea viscosa (L.) Jacq.

Common name
shahs

Description
Erect evergreen shrub or small tree, resinous when young; up to 2 m.

Leaves Alternate, dark green, glossy, lance-shaped with entire margins and an acute tip.

Flowers Inconspicuous red/green, clustered in the leaf axils. Unisexual.

Fruit/seed Showy capsule with deep red or pale yellow papery wings. Seeds black.

Flowering period
January–May.

Habitat
The edges of dry woodland and on rocky slopes; altitude from 150–2,500 m.

Distribution
Native to Australia, now pan-tropical. Commonly cultivated as an ornamental.

Note
Named after Rembert Dodoens, a sixteenth-century Dutch herbalist.

Family Scrophulariaceae (Foxglove or Figwort Family)

Name

Campylanthus pungens O.Schwartz

Description
A small ascending shrub with semi-succulent stems; up to 1 m.

Leaves Usually leafless.

Flowers Dark pink with a double bend in the corolla tube.

Fruit/seed Capsule with many black seeds.

Flowering period
September–November.

Habitat
Dry hillsides in Dhofar; altitude up to 1,600 m.

Distribution
Regional endemic to Dhofar, Yemen and Saudi Arabia.

Note
The name derives from the Greek meaning 'bent flower' and refers to the corolla. There are three additional *Campylanthus* species in Oman, all have the characteristic bent corolla and are endemic.

Family Scrophulariaceae (Foxglove or Figwort Family)

Name

Schweinfurthia imbricata A.G. Mill., M.Short & D.A.Sutton

Description

Prostrate with young shoots covered in white hair; up to 15 cm.

Leaves Blue/grey, overlapping, slightly succulent, round with entire margins.

Flowers Pale pink or cream flowers appear between the leaves.

Fruit/seed Dark brown, round capsule with a pale brown papery covering and oblong seeds.

Flowering period

February–May.

Habitat

Gravel plains, wadis, rocky slopes and along roadsides; altitude from 60–650 m.

Distribution

Regional endemic to northern Oman and UAE.

Uses

Crushed leaves are used to treat snakebite.

Family Scrophulariaceae (Foxglove or Figwort Family)

Name

Striga gesnerioides (Willd.) Vatke ex Engl.

Description An erect dark reddish herb. Parasitic on the roots of other plants; up to 45 cm.

Leaves Inconspicuous, scale like, reddish.

Flowers Pink, in clusters along the stems.

Fruit/seed An oblong capsule with numerous minute seeds.

Flowering period
August–November.

Habitat
In monsoon areas of the Dhofar escarpment; altitude from 200–1,200 m.

Distribution
Widespread in the Old World tropics.

Uses
As a dye for cosmetics.

Note
There are two further species of *Striga* in Dhofar.

Family Tamaricaceae (Tamarisk Family)

Name

Tamarix aphylla H.Karst.

Common names
Tamarisk, athl

Description
A much branched shrub or small tree with reddish bark; up to 6 m.

Leaves Alternate, minute, scale-like, grey-green with salt secreting glands.

Flowers Tiny pink sprays spiral along the stem.

Fruit/seed Brown, many seeded capsule. Seeds with an apical tuft of hair.

Flowering period
October–April.

Habitat
Wadi banks throughout the country; altitude up to 1,000 m.

Distribution
Mediterranean, Africa, Arabia and Pakistan. Frequently grown as an ornamental.

Uses Wood is termite resistant and used for building.

Note
Deep rooted and tolerant of salt, useful in stabilising dunes. Several species of *Tamarix* are found in Oman.

Family	Verbenaceae (Vervain Family)

Name

Phyla nodiflora (L.) Greene

Description
Prostrate carpet-forming perennial, rooting at the nodes; up to 30 cm.

Leaves Opposite, rounded, tapering towards the base with serrated margins.

Flowers Pink or white with yellow centres, congested on an erect spike.

Fruit/seed Tiny, yellow ovoid capsules.

Flowering period
All year.

Habitat
Common in wet areas with standing water, a weed of cultivation; altitude up to 640 m.

Distribution
Pan-tropical, sometimes cultivated as ground cover.

Family Zygophyllaceae (Creosote bush or Lignum-vitae Family)

Name

Fagonia indica Burm.f.

Common name
shukaa

Description
Low growing woody perennial with branching square and grooved stems, paired spines at the leaf nodes; up to 80 cm.

Leaves Opposite, small, oval, bluish, with entire margins, trifoliate near the base, upper leaflets single.

Flowers Solitary, fragrant, pink with five free petals which may turn back, prominent yellow stamens.

Fruit/seed Onion-shaped capsule with five one-seeded segments, covered in hair.

Flowering period
Most of the year.

Habitat
Commonly found in open areas, along roadsides, on mountains and in wadis; altitude up to 2,000 m.

Distribution
Throughout the Old World tropics.

Uses
Medicinal uses reported, including treatment for sexually transmitted diseases.

Note
There are five further species of *Fagonia* in Oman, some do not have spines. Named after G. C. Fagin, a French botanist (1638–1718).

Yellow

and orange flowers

Family Acanthaceae (Acanthus Family)

Name

Barleria proxima Lindau

Description

Dense low growing bush with prominent spines; up to 25 cm.

Leaves Opposite, grey/green, oval with entire margins.

Flowers Bright orange with four large petals above and a smaller, fifth petal below the stamens.

Fruit/seed Capsule with a long beak containing two seeds.

Flowering period

Most of the year.

Habitat

Locally common in wadis and rocky slopes in Dhofar; altitude up to 1,300 m.

Distribution

E Africa and southern Arabia.

Family Acanthaceae (Acanthus Family)

Name

Justicia odora (Forssk.) Lam.

Description

A woody based perennial; up to 30 cm.

Leaves Opposite, often appearing whorled, overall ovate with an indented tip and hairs along the margins.

Flowers Bright yellow, with a lobed lower lip, usually in pairs arising at the leaf nodes.

Fruit/seed A hairy capsule with flattened seeds.

Flowering period

September–November.

Habitat

Dry zone of Dhofar, beyond the influence of the monsoon; altitude up to 1,500 m.

Distribution

Africa and tropical Arabia.

Family Aizoaceae (Carpet-weed or New Zealand Spinach Family)

Name

Aizoon canariense L.

Common name
Purslane-leaved aizoon, shuheymat ard

Description
Annual or short-lived woody perennial with zig-zagging branches radiating out from the centre; up to 20 cm.

Leaves Alternate, bright green, oval with entire margins.

Flowers Tiny, borne in forks of branches surrounded by five hairy sepals.

Fruit/seed Star-shaped red capsule.

Flowering period
Most of the year.

Habitat
Common on sandy and gravel areas; altitude up to 1,800 m.

Distribution
N Africa, Canary Islands, Mediterranean, Palestine, Arabia, SW Asia.

Uses
Leaves eaten in salads.

Family Apocynaceae: Asclepiadoideae (Oleander or Milkweed Family)

Name

Caralluma quadrangula (Forssk.) N.E.Br.

Description

Succulent, stems square, erect with white leaf scars along the ridges, forming clumps; up to 50 cm high.

Leaves Absent.

Flowers Five triangular lobes with a darker centre.

Fruit/seed Paired, green, erect follicles. Seeds pale brown with tufts of hair.

Flowering period

August–November.

Habitat

Dry rocky slopes and the high plateau of Dhofar; altitude up to 1,200 m.

Distribution

Regional endemic to southern Oman, Yemen and Saudi Arabia.

Uses

The whole plant is edible.

Family Apocynaceae: Asclepiadoideae (Oleander or Milkweed Family)

Name

Echidnopsis scutellata (Defl.) A.Berger

Description

An apparently leafless plant with succulent cylindrical stems, often prostrate.

Leaves Very small, rudimentary near apex of the stem, soon falling.

Flowers Borne directly on the stem with five yellow petals.

Fruit/seed Paired green, erect follicles.

Flowering period

August–November.

Habitat

Rocky slopes and cliffs in the drier parts of Dhofar; altitude from 700–1,500 m.

Distribution

Endemic to Dhofar and Yemen.

Family Apocynaceae: Asclepiadoideae (Oleander or Milkweed Family)

Name

Glossonema varians Benth. ex Hook.f.

Common name
itr

Description
Annual or perennial herb branched from the base, covered in white hair with white latex; up to 30 cm.

Leaves Opposite, somewhat fleshy, hairy, broadly round with wavy margins.

Flowers Pale yellow with a red centre, arising from the leaf axils.

Fruit/seed Ellipsoid with soft spines hardening with age. Seeds brown, flattened with a tuft of white hair.

Flowering period
After rain.

Habitat Common in gravel plains; altitude up to 1,300 m.

Distribution
Arabia and east to Pakistan and India.

Uses
Fruit is a popular food, foliage used for animal fodder.

Family Apocynaceae: Asclepiadoideae (Oleander or Milkweed Family)

Name

Leptadenia pyrotechnica (Forssk.) Decne.

Common name
moorakh

Description
Bushy perennial with smooth cylindrical stems; up to 3 m.

Leaves Lance-shaped, soon falling, leaving the stems naked.

Flowers Numerous, small, borne at the stem nodes, fragrant at dusk.

Fruit/seed A long narrow follicle. Seeds have long hairs.

Flowering period
November–June.

Habitat
Dry, sandy areas; altitude up to 300 m.

Distribution
N Africa, Arabia and east to Pakistan and India.

Uses
Buds are edible and much appreciated. Dried hair from the seeds is used for kindling.

Note
Small mammals shelter under this plant in the day.

Family Apocynaceae: Asclepiadoideae (Oleander or Milkweed Family)

Name

Pentatropis nivalis (J.F.Gmel.) D.V.Field & J.R.I.Wood

Common name
schuntob

Description
Twining perennial vine; climbs up to 5 m.

Leaves Pale green, opposite, oblong with entire margins.

Flowers Five, pale yellow, spiral petals, giving a spidery appearance, in small clusters at the leaf nodes.

Fruit/seed Narrow beaked follicle. Seeds small, flattened, with a tuft of hair at the top.

Flowering period
Most of the year.

Habitat
Generally in poor sandy and silty soils, often covering *Acacia* bushes, never free standing; altitude up to 900 m.

Distribution
Widely distributed throughout the Old World tropics.

Uses
The young pods are edible. Roots used to treat gonorrhoea in India.

Family Apocynaceae: Asclepiadoideae (Oleander or Milkweed Family)

Name

Periploca visciformis (Vatke) K.Schum.

Common name
handabub

Description
Upright bushy plant with branched, leafless green stems; up to 1 m.

Leaves Small, soon falling.

Flowers Pale yellow, reflexed petals with a darker stripe down the centre, borne directly on the stems.

Fruit/seed Paired or single cylindrical follicles containing many red/brown seeds with a tuft of hair.

Flowering period
November–April.

Habitat
Mainly in the higher, drier areas of Dhofar; altitude up to 1,600 m.

Distribution
Tropical NE Africa and southern Arabia.

Uses
Fodder plant; maintains its succulent stems during most of the dry season.

Note
Favoured by Ibex.

Family Asphodelaceae (Aloe Family)

Name

Aloe vera L.

Common names
Barbados aloe, siqal

Description
Succulent perennial usually found growing in clumps; up to 1.5 m.

Leaves Grey, forming a basal rosette, lance shaped blades up to 60 cm long with spiny margins.

Flowers Numerous, yellow/red on a long stalk arising from the basal rosette.

Fruit/seed Oval, many seeded capsule.

Flowering period
Most of the year.

Habitat
Found on rocky slopes and in wadis, probably escaped from cultivation; altitude up to 1,500 m.

Distribution
Native to N Africa, now pan-tropical. Grown as an ornamental in towns.

Uses
Widely used in cosmetics and medicinally.

Family Asteraceae (Daisy or Sunflower Family)

Name

Centaurea pseudosinaica Czerep.

Description
An erect annual with bristly winged stems; up to 40 cm.

Leaves Alternate, lance-shaped with wavy or serrated margins.

Flowers Bright yellow compact flower heads with long pale yellow spines below.

Fruit/seed Single-seeded achene attached to a tuft of hair.

Flowering period
February–April.

Habitat
Common in foothills after rain; altitude up to 500 m.

Distribution
Arabia, Iran and Iraq.

Family Asteraceae (Daisy or Sunflower Family)

Name

Euryops arabicus Steud. ex Jaub. & Spach

Common names
hanqalan, hayqalan

Description
Small, spreading shrub; up to 80 cm.

Leaves Dark green, crowded towards the top of the branches, linear, pointing upwards.

Flowers Single yellow flower heads with a sweet scent.

Fruit/seed Brown achenes with a tuft of white hair.

Flowering period
Most of the year.

Habitat
Common in the northern mountains; altitude from 1,000–2,800 m.

Distribution
NE Africa and Arabia.

Family Asteraceae (Daisy or Sunflower Family)

Name

Helichrysum glumaceum DC.

Description
A small, greyish perennial with spreading stems; up to 25 cm.

Leaves Alternate, pale grey, concentrated at the base of the plant, edges curled inwards.

Flowers Tight clusters of pale yellow flowers in round heads, surrounded by white bracts; never open fully.

Fruit/seed Brown achenes with glands and tufts of hair on the top.

Flowering period
February–April.

Habitat On rocky hillsides; altitude up to 1,000 m.

Distribution
Tropical Africa and Arabia.

Family Asteraceae (Daisy or Sunflower Family)

Name

Iphiona aucheri (Boiss.) Anderb.

Common name
khassya

Description
Compact perennial herb; up to 80 cm.

Leaves Alternate, bright green, slightly succulent, broadly triangular, tapering towards the short stalk, margins entire except at the top where they are roughly serrated.

Flowers Bright yellow, terminal flower heads.

Fruit/seed Brown achene.

Flowering period
January–May.

Habitat
Common along roadsides and wadi flows in northern Oman; altitude up to 600 m.

Distribution
UAE and SW Asia.

Note
Very poisonous, and may kill camels.
I. scabra and *I. horrida* are similar but have thin needle-like leaves and smaller flowers.

Family Asteraceae (Daisy or Sunflower Family)

Name

Launaea spp.

Description

General appearance ranges from compact, rounded mounds, to tall open plants; up to 60 cm.

Leaves Basal in taller species, serrated, often deeply divided, often greyish.

Flowers Usually solitary at the end of slender stems.

Fruit Achenes, with copious white hair.

Flowering period

After rain.

Habitat

Disturbed areas, edges of roads and plantations; altitude up to 2,500 m.

Distribution

Widely distributed from the Canary Islands to S Africa and E Asia. There are twelve species of *Launaea* in Oman.

Name

Launaea castanosperma F.G.Davies

Annual with long, thin, branching stems; up to 50 cm. Leaves with finely dissected margins form a basal rosette. Common in Dhofar.

Family Asteraceae (Daisy or Sunflower Family)

Name

Launaea bornmuelleri (Hausskn. ex Bornm.) Bornm.

Compact cushion-forming plant with zigzagging branches. Found on rocky ground up to 2,000 m, less common in lowlands.

Name

Launaea intybacea (Jacq.) Beauv.

An annual with an open aspect. Leaves deeply segmented. Flowers pale yellow. Common as a weed of cultivation in lowland areas.

Family Asteraceae (Daisy or Sunflower Family)

Name

Pulicaria argyrophylla Gamal-Eldin

Description
A cushion-forming perennial; up to 50 cm.

Leaves Alternate, grey/green slightly succulent with entire margins.

Flowers Single, round heads with open yellow flowers.

Fruit/seed Brown achene with tufts of hair.

Flowering period
Most of the year.

Habitat
Rocky and sandy places and along roadsides in southern Oman; altitude up to 1,400 m.

Distribution
NE Africa and southern Arabia.

Note
There are several species of *Pulicaria* in Oman.

Family Asteraceae (Daisy or Sunflower Family)

Name

Pulicaria glutinosa Jaub. & Spach

Description

A much branched woody perennial; up to 1 m.

Leaves Alternate, narrow, point upwards, slightly glossy with an aromatic smell.

Flowers Terminal on leafless stalks.

Fruit/seed Brown achene with tufts of hair at one end.

Flowering period

January–May.

Habitat

Gravel plains, mountain slopes, wadis and foothills; altitude up to 2,000 m.

Distribution

Endemic to Arabia.

Note

The genus *Pulicaria* commemorates a French Abbot, Noel Antoine Pluche (1685–1761).

Family Asteraceae (Daisy or Sunflower Family)

Name

Reichardia tingitana (L.) Roth

Description Annual herb with short stems; up to 20 cm.

Leaves Lower leaves form a basal rosette, blue/green, oblong with deeply serrated margins. Upper leaves are small, with entire margins and clasp the stem.

Flowers Bright yellow with dark centres, solitary on a thick, erect stalk.

Fruit/seed Achene with tufts of hair.

Flowering period
February–April.

Habitat
In wadis, gravel plains and on rocky slopes in northern Oman; altitude up to 1,200 m.

Distribution
N and E Africa, Mediterranean, Middle East and India.

Uses
Dried leaves are used in the UAE to treat constipation and colic.

Family Boraginaceae (Borage or Forget-me-not Family)

Name

Arnebia hispidissima (Lehm.) DC.

Common names
Arabian primrose, habrak

Description
An erect annual with branched stems covered in stiff hair; up to 20 cm.

Leaves Alternate, greyish, lance-shaped, hairy.

Flowers Single, yellow flowers with petals slightly curved backwards.

Fruit/seed Four glossy nutlets.

Flowering period
December–May.

Habitat
Common after rains in sand and gravel areas; altitude up to 400 m.

Distribution
N and NE Africa, Arabia, Middle East, Iran and Pakistan.

Uses
Has well documented antimicrobial properties. Dye from the roots is used as a cosmetic in the UAE.

Family Brassicaceae (Cabbage or Mustard Family)

Name

Diplotaxis harra (Forssk.) Boiss.

Common name
khawshyan

Description
Erect annual or perennial herb, hairy; up to 60 cm.

Leaves Forms a basal rosette, oval with serrated margins, at lower altitudes they are hairy but nearly smooth at high altitudes.

Flowers Bright yellow with four petals, borne at the end of long erect stalks.

Fruit/seed Long and narrow with a beak hanging down. Seeds round, red/brown.

Flowering period
Most of the year.

Habitat
Common in gravel and sandy areas; altitude up to 2,300 m.

Distribution
N Africa, Arabia and SW Asia.

Note
Attracts the Krueperi small white butterfly (*Artogeia krueperi*).

Family Brassicaceae (Cabbage or Mustard Family)

Name

Sisymbrium erysimoides Desf.

Common name
maharraqa

Description
Erect annual, sparsely covered in white hairs; up to 70 cm.

Leaves Alternate, mid-green with a short stalk and serrated margins, lower leaves are lobed with irregularly toothed margins and pointed tips, upper leaves are smaller.

Flowers Small, occur in dense terminal sprays.

Fruit/seed Long and narrow on long thick stalks. Seeds orange/brown.

Flowering period
January–May.

Habitat
Gravel and rocky areas and as a weed in plantations; altitude up to 2,300 m.

Distribution
Mediterranean, Africa, Arabia and SW Asia.

Note
S. irio differs in the shape of the fruit, which is more compact.

Family Capparaceae (Caper Family)

Name

Cadaba farinosa Forssk.

Description

Small, densely branched shrub with a white powdery covering; up to 2 m.

Leaves Alternate, grey/green, oval with entire margins, covered in white powder.

Flowers Four pale yellow petals above four green/brown sepals with long protruding stamens.

Fruit/seed Cylindrical, many seeded with a white powdery covering.

Flowering period

Most of the year.

Habitat

Common in dry areas on rocky slopes and gravel; altitude up to 1,200 m.

Distribution

Tropical Africa to Pakistan and India.

Uses

Juice from the leaves is used as eye drops.

Note

Attracts caper white (*Anaphaeis aurota*) and zebra white (*Pinacopteryx eriphia*) butterflies.

Family Caprifoliaceae (Honeysuckle Family)

Name

Lonicera aucheri Jaub. & Spach

Common name

kanum

Description

Dense, woody shrub with spreading branches and red stems; up to 2 m.

Leaves Opposite, oval with entire margins.

Flowers Mass of sweet scented yellow flowers.

Fruit/seed Fleshy berries turning red as they mature.

Flowering period

March–September.

Habitat

Rocky slopes of the Jabal Al Akhdar range; altitude above 2,000 m.

Distribution

Northern Oman, Iran and Iraq.

Family Chenopodiaceae (Goosefoot, Beetroot or Spinach Family)

Name

Sevada schimperi Moq.

Description
Small spreading shrub with a woody base; up to 40 cm.

Leaves Alternate, grey/green, succulent, cylindrical with a tuft of hair in the axil.

Flowers Yellow, forming small clusters.

Fruit/seed A ribbed tube with an acute tip.

Flowering period
September–November.

Habitat
Sandy and saline areas of Dhofar and central Oman; altitude up to 100 m.

Distribution
Tropical E Africa, Egypt and southern Arabia.

Family Cistaceae (Rockrose Family)

Name

Helianthemum citrinum Ghaz.

Description

A low growing perennial; up to 20 cm.

Leaves Opposite, greyish, oval, hairy with entire margins.

Flowers Bright yellow with reflexed petals, on one side of the stalk.

Fruit/seed An ovoid, brown capsule.

Flowering period

September–October.

Habitat

Rocky ground in Dhofar; altitude from 100–1,250 m.

Distribution

Endemic to Dhofar and SE Yemen.

Note

There are three further species of *Helianthemum* in Oman, mainly found in the northern mountains.

Family Cleomaceae (Spider Flower or African Cabbage Family)

Name

Cleome austroarabica D.F.Chamb. & Lamond

Description Erect, dwarf perennial with a lemony scent; up to 1 m.

Leaves Alternate, oval to round with entire margins, covered in sticky glands.

Flowers Pale yellow petals folded back with prominent yellow stamens.

Fruit/seed Oblong, erect capsule.

Flowering period Most of the year.

Habitat Rocky and limestone areas; altitude up to 1,750 m.

Distribution Endemic to southern Arabia.

Uses Sweet smelling leaves are crushed and rubbed over the body as a deodorant.

Note
There are 11 *Cleome* species in Oman.

Family Cleomaceae (Spider Flower or African Cabbage Family)

Name

Cleome rupicola Vicary

Description

A woody perennial with slender stems, hairy, strongly aromatic; up to 80 cm.

Leaves Alternate, glandular, oval with entire margins.

Flowers Dark yellow/orange with brown veins, four petals curved backwards, stamens prominent, several flowers borne on a terminal spike.

Fruit/seed Straight capsule hangs down below the flowers. Seeds woolly.

Flowering period

February–April.

Habitat

Common in sand and gravel areas; altitude up to 1,500 m.

Distribution

Arabia, Iran and Pakistan.

Uses

Extract of the whole plant used to treat eye problems in UAE.

Family Cleomaceae (Spider Flower or African Cabbage Family)

Name

Dipterygium glaucum Decne.

Common name
alga

Description
Erect perennial shrub with yellowish/green branched stems; up to 80 cm.

Leaves Alternate, linear, tend to fall soon, leaving the stems appearing naked.

Flowers Yellow with four petals, borne on short stalks, in lax clusters.

Fruit/seed Solitary, light green to beige, round and flat with a winged surface.

Flowering period
January–April.

Habitat
On sand and gravel, often along roadsides; altitude up to 500 m.

Distribution
NE Africa, Arabia and east to Pakistan.

Family Convolvulaceae (Morning-glory or Bindweed Family)

Name

Ipomoea obscura (L.) Ker Gawl.

Description

Perennial climber with twining stems that can reach 2 m.

Leaves Alternate, mid-green, heart-shaped with entire margins.

Flowers Solitary, bright yellow to orange with a purple centre.

Fruit/seed A round capsule with black hairy seeds.

Flowering period

Most of the year.

Habitat

Common on the wet escarpment of Dhofar; altitude up to 1,000 m.

Distribution

Native to the Old World Tropics.

Family Convolvulaceae (Morning-glory or Bindweed Family)

Name

Merremia somalensis Hallier f.

Description

Annual herb with long prostrate stems, may have fine hairs; up to 1 m.

Leaves Alternate, oblong with serrated edges.

Flowers Dull yellow with a red centre, usually solitary but may be in pairs.

Fruit/seed A capsule with hairy sepals. Seeds black and flattened.

Flowering period

August–October.

Habitat

The monsoon escarpment of Dhofar; altitude up to 700 m.

Distribution

Somalia, southern Oman and Yemen.

Family Cucurbitaceae (Pumpkin, Gourd or Cucumber Family)

Name

Citrullus colocynthis (L.) Schrad.

Common names
Desert squash, bitter squash, handal

Description
Trailing perennial herb; stems up to 2 m long.

Leaves Alternate, grey/green, overall triangular in shape, deeply lobed with entire margins, tendrils appearing from the leaf nodes.

Flowers Solitary, arising from the leaf axils.

Fruit/seed Large round, smooth, turning from green with yellow stripes to yellow as they ripen. Seeds are embedded in bitter white pulp.

Flowering period
January–April.

Habitat
Sandy and gravel areas; altitude up to 1,250 m.

Distribution
Tropical Africa and Arabia to SW India.

Uses
Numerous medicinal properties are claimed, including use as a laxative and to relieve the pain of scorpion stings.

Family Cucurbitaceae (Pumpkin, Gourd or Cucumber Family)

Name

Corallocarpus epigaeus (Roettl.) Hook.f. ex Clarke

Description
Climbing vine with papery bark; up to 4 m.

Leaves Alternate, oval with 3–5 deep lobes and entire or slightly serrated margins.

Flowers Erect, borne on a long stalk in the leaf axils.

Fruit/seed Bright red with a green cup at the base.

Flowering period
Most of the year.

Habitat
Dry gravel and rocky areas; altitude from 100–1,200 m.

Distribution
NE and W Africa, Arabia, Pakistan and India.

Family Cucurbitaceae (Pumpkin, Gourd or Cucumber Family)

Name

Cucumis prophetarum L.

Description
Climbing annual; stems trailing up to 2 m.

Leaves Alternate, grey/green, on long stalks with 3–5 deep lobes and wavy margins.

Flowers Solitary or in small clusters, with five petals, borne in the leaf axils.

Fruit/seed Young fruit are green and yellow, turning completely yellow with soft spines when mature.

Flowering period
Most of the year.

Habitat
Locally common in wadis, sand and gravel areas; altitude up to 1,250 m.

Distribution
Tropical Africa, Arabia and SW Asia.

Family Cucurbitaceae (Pumpkin, Gourd or Cucumber Family)

Name

Cucumis sativus L.

Common names
Wild cucumber, khigar

Description Vigorous climbing annual, stems ribbed; up to 5 m.

Leaves Alternate, bright green, broadly heart shaped with an acute tip and irregular margins, tendrils arise from the leaf axils.

Flowers Bright yellow with five petals which turn to face the sun, even when cloudy.

Fruit/seed Large, cylindrical, green, covered in short bristles.

Flowering period
July–October.

Habitat
Common on the wooded hillsides of the escarpment in Dhofar; altitude from 100–900 m.

Distribution
Throughout the tropics, cultivated in cooler climates.

Note
The wild cucumber is very bitter and not edible.

Family Ephedraceae (Gymnosperm)

Name

Ephedra pachyclada Boiss.

Common names
ansab, ansaran

Description
Small, dense shrub with stiff, erect branches; up to 100 cm.

Leaves Opposite, unobtrusive, scale-like.

Flowers Yellow cones, borne in clusters in the leaf axils.

Fruit/seed Ripe cones are red and fleshy.

Flowering period
May–June.

Habitat
Rocky slopes of the northern mountains, in full sun; altitude above 1,250 m.

Distribution
Arabia and SW Asia.

Uses
Reported as a treatment for symptoms of asthma. Fruit is edible.

Family Euphorbiaceae (Spurge Family)

Name

Euphorbia cactus Ehrenb. ex Boiss.

Description

An erect spiny shrub with dark green stems that have zigzagging angles with spines on the ridges; up to 1 m.

Leaves Inconspicuous and deciduous.

Flowers Yellow, between the spines on the angles at the top of the stems.

Fruit/seed A three-lobed capsule. Seeds smooth and grey.

Flowering period

August–November.

Habitat

Rocky areas and cliffs on the escarpment in Dhofar; altitude from 300–850 m.

Distribution

Ethiopia and SW Arabia.

Uses

Fodder for camels in the dry season.

Note

The genus is named after Euphorbus the Greek physician to King Juba of Mauritania. Sap is caustic.

Family Leguminosae (Pea, Bean or Mimosa Family)

Name

Crotalaria aegyptiaca Benth.

Common name
nizaar

Description
Bushy perennial with stiff grey/green ribbed branches, forked at right angles and slightly hairy; up to 45 cm.

Leaves Alternate, very sparse, oval.

Flowers Single, pea-like, appear at regular intervals along the upper branches.

Fruit/seed Greyish, soft pod with 1–4 yellow/brown seeds.

Flowering period
January–April.

Habitat
Sand, gravel and rocky areas; altitude up to 850 m.

Distribution
NE Africa, Arabia, Israel, Palestine and Jordan.

Note
Heavily grazed by livestock.

Family Leguminosae (Pea, Bean or Mimosa Family)

Name

Crotalaria retusa L.

Description

Bushy perennial with erect, unbranched stems; up to 150 cm.

Leaves Alternate, oblong with entire margins.

Flowers Pea-like with red veins, occur along the stems.

Fruit/seed Plump reddish-brown pods with many seeds, borne along the branches.

Flowering period

Most of the year.

Habitat

A weed of cultivation and in the grasslands of Dhofar; altitude up to 1,000 m.

Distribution

Pan-tropical.

Note

Toxic to sheep.

Family Leguminosae (Pea, Bean or Mimosa Family)

Name

Melilotus indicus (L.) All.

Common names
Scented trefoil, Indian melilot

Description
Small annual with branched stems; up to 50 cm.

Leaves Alternate, trifoliate with serrated margins, tips blunt or indented.

Flowers Small, in clusters at the top of stems.

Fruit/seed Round pods with ridges.

Flowering period
February–May.

Habitat
Weed in cultivated areas; altitude up to 2,300 m.

Distribution
Mediterranean, Middle East to India; introduced in warm temperate regions.

Family Leguminosae (Pea, Bean or Mimosa Family)

Name

Pseudolotus villosus (Blatt. & Hallb.) Ali & D.D.Sololoff

Description
Small, prostrate annual, hairy; up to 15 cm.

Leaves Opposite, crowded on stems, greyish, oblong, hairy with entire margins.

Flowers Single or paired, yellow with red markings, emerging between the leaves.

Fruit/seed A smooth brown pod.

Flowering period
February–April.

Habitat
Rocky hillsides in northern Oman; altitude up to 1,500 m.

Distribution
Arabia, Iran and Pakistan.

Family Leguminosae (Pea, Bean or Mimosa Family)

Name

Senna holosericea (Fresen.) Greuter

Common names
ishriq, kharkhash

Description
Erect, woody, hairy perennial; up to 60 cm.

Leaves Alternate, greyish, compound with 4–7 paired leaflets with acute tips, hairy.

Flowers Up to 15 flowers clustered on erect terminal or axial stalks.

Fruit/seed Pods wide and flat, may have a slight curve, green, turning black as they mature.

Flowering period
Most of the year.

Habitat
In compacted sand and gravel, and as a weed of cultivation; altitude up to 800 m.

Distribution
N Africa, Arabia, SW Asia.

Note
S. italica is very similar but with a ridge on the crest of the fruit.

Family Malpighiaceae

Name

Acridocarpus orientalis A.Juss.

Common name
qafas

Description
A large shrub with whitish bark; up to 3 m.

Leaves Alternate, elliptic with entire margins and an acute tip, young leaves have reddish hair along the veins, older leaves are smooth but with prominent veins.

Flowers Five free petals, in dense clusters at the end of the branches.

Fruit/seed Paired with a single seed surrounded by a large pale brown wing.

Flowering period
Most of the year.

Habitat
Gravel wadis and rocky slopes; altitude from 300–1,500 m.

Distribution
E Africa and Arabia.

Uses
Crushed leaves are mixed with oil to reduce swellings and muscle or joint pain. Used as a yellow dye.

Note
Attracts the giant skipper butterfly (*Coeliades anchises*).

Family Malvaceae (Mallow, Cotton or Hollyhock Family)

Name

Abutilon pannosum (G.Forst) Schltdl.

Common name
munnaqa

Description
Erect perennial, hairy; up to 2 m.

Leaves Alternate, grey/green, heart-shaped with serrated margins and silvery hair.

Flowers Single with five free petals, opening in the late afternoon.

Fruit/seed In densely packed sections. Seeds brown, hairy.

Flowering period
Most of the year.

Habitat
Common along roadsides and in disturbed areas; altitude up to 400 m.

Distribution
Tropical Africa, Arabia, Middle East and SW Asia.

Note
There are six species of *Abutilon* in Oman.

Family Oxalidaceae (Wood-sorrel, Bermuda Buttercup, Oca or Starfruit Family)

Name

Oxalis corniculata L.

Common name
mahummad

Description
Spreading prostrate annual or perennial rooting at the nodes; up to 30 cm.

Leaves Alternate, compound with three heart-shaped leaflets with entire margins, hairy below and along the margins.

Flowers Arise in the leaf axils in sprays of 1–6 flowers.

Fruit/seed Cylindrical, grooved capsule with brown seeds.

Flowering period
February–May.

Habitat
Shady places near water and as a weed of cultivation; altitude up to 2,300 m.

Distribution
Cosmopolitan weed.

Note
Toxic to livestock.

Family Papaveraceae (Poppy Family)

Name

Argemone mexicana L.

Common names
Mexican poppy, Prickly poppy, ghadrect

Description
Erect, spiny annual, stems with yellow latex; up to 1 m.

Leaves Alternate, blue/green with a distinctive white stripe down the centre and along the veins, deeply lobed with sharp, spiny margins.

Flowers Bright yellow, papery, terminal.

Fruit/seed Oblong, erect, ribbed capsule with sharp prickles. Seeds black.

Flowering period
Most of the year.

Habitat
Disturbed, open areas and wadis in Dhofar; altitude up to 900 m.

Distribution
Native to Mexico, now an invasive pan-tropical weed.

Uses
Various medicinal uses are reported including treating eye infections.

Note
Toxic.

Family Portulacaceae (Purslane or Lewisia Family)

Name

Portulaca oleracea L.

Common names
barbir, baklah, nijlah, rigla

Description
Spreading, prostrate annual, often tinged red; up to 30 cm.

Leaves Oval, fleshy, with entire margins.

Flowers Yellow with a notch at the top, open in the mornings only.

Fruit/seed An oval capsule with many tiny seeds.

Flowering period
Most of the year.

Habitat
Edges of plantations, damp disturbed soil; altitude up to 2,000 m.

Distribution
Cosmopolitan.

Uses
Edible, used as a cure for diarrhoea and as a diuretic.

Family Ranunculaceae (Buttercup or Monkshood Family)

Name

Ranunculus muricatus L.

Common name
Spinyfruit buttercup

Description
Erect annual herb with slender stems; up to 35 cm.

Leaves Opposite, overall round with three lobes and deeply serrated margins, on long stalks.

Flowers Five free, glossy petals, solitary on long stalks.

Fruit/seed Bright green with a long beak, in clusters on a stalk.

Flowering period
February–May.

Habitat
Cultivated and damp shady areas in northern Oman; altitude from 300–2,200 m.

Distribution
S Europe, N Africa, Arabia and SW Asia.

Family Resedaceae (Mignonette or Weld Family)

Name

Ochradenus arabicus Chaudhary, Hillc. & A.G. Mill.

Common names
qurliya, asmat, hibab

Description
Shrub, forming dense clumps of branching green stems ending in spines; up to 1 m.

Leaves Small, narrow, fall early giving a leafless appearance.

Flowers Bright yellow, on terminal spikes. During the flowering period they appear to cover the bush in colour.

Fruit/seed Ovoid capsule, clustered in groups, turning from green to pale yellow as they mature. Seeds black and glossy.

Flowering period
Most of the year.

Habitat
Locally common on open rocky hillsides; altitude up to 2,000 m.

Distribution
Regional endemic to southern Arabia.

Note
O. baccatus is similar and distinguished by white fruit which turn red when dry, *O. aucheri* has paler flowers at intervals along the stem. Attracts the desert white butterfly (*Pontia glauconome*).

Family Rutaceae (Rue or Citrus Family)

Name

Haplophyllum tuberculatum (Forssk.) A.Juss.

Common names
al taies, tafar al tays

Description
Erect woody perennial covered in small raised glands with a strong, unpleasant odour; up to 60 cm.

Leaves Alternate, vary in shape and margins, but are overall oval with a tapering base.

Flowers Bright yellow with five free petals, borne in terminal clusters on short stalks.

Fruit/seed A round, five-lobed capsule. Seeds kidney-shaped.

Flowering period
Most of the year.

Habitat
Gravel plains, rocky slopes and along roadsides; altitude up to 1,500 m.

Distribution
N and E Africa, the Middle East and east to Pakistan.

Uses
The leaves and sap have a variety of medicinal uses but are said to be poisonous if consumed in large quantities.

Note
The unpleasant smell is thought to deter insects, therefore bunches are hung at the entrance to houses. Attracts the swallow tail butterfly (*Papilio machaon*).

Family Scrophulariaceae (Foxglove or Figwort Family)

Name

Alectra parasitica A.Rich.

Description

An erect branched parasite; up to 30 cm.

Leaves Opposite, scale-like, slightly thickened with an acute tip.

Flowers Dense spikes of small flowers with purple veins, five petals with leaf-like bracts.

Fruit/seed Round capsule with many minute seeds.

Flowering period

August–November.

Habitat

Damp areas within the monsoon area of Dhofar; altitude from 600–1,100 m.

Distribution

Tropical Africa, Ethiopia, Somalia and southern Arabia.

Family Scrophulariaceae (Foxglove or Figwort Family)

Name

Cistanche phelypaea (L.) Cout.

Common names
basul, dhamin

Description
Parasite, often growing on species of *Atriplex*, stem upright, almost colourless, obscured by early flowers that have died; up to 70 cm.

Leaves Scale-like, unobtrusive.

Flowers Bright yellow new flowers cluster around the top of the spike.

Fruit/seed Ovoid black capsule. Seeds minute.

Flowering period
After rains.

Habitat
Sandy and coastal regions; altitude up to 350 m.

Distribution
N and E Africa, the Mediterranean, Arabia and east to India.

Family Scrophulariaceae (Foxglove or Figwort Family)

Name

Schweinfurthia papilionacea Boiss.

Description

An erect annual, branched from the base; up to 60 cm.

Leaves Alternate, round, leathery, slightly fleshy with entire margins.

Flowers Two-lipped, with pink veins, emerging from between the leaves.

Fruit/seed A round capsule with many black seeds.

Flowering period

February–May.

Habitat

Wadi beds and gravel plains in northern and central Oman; altitude up to 600 m.

Distribution

Arabia and east to India and Pakistan.

Family Sparrmanniaceae (Jute, Sparrmannia or Grewia Family)

Name

Corchorus depressus (L.) Stocks

Common name
matara

Description
Prostrate woody perennial forming a dense mat; up to 30 cm.

Leaves Alternate, dark green, oval with undulating margins.

Flowers Bright yellow with five petals that open in the afternoon.

Fruit/seed A woody, cylindrical capsule.

Flowering period
Most of the year.

Habitat
Common in wadis, gravel and sandy plains; altitude up to 500 m.

Distribution
N Africa, Arabia and east to India.

Family Sparrmanniaceae (Jute, Sparrmannia or Grewia Family)

Name

Triumfetta pentandra A. Rich. ex Guill.

Description
An erect, branched, hairy annual; up to 60 cm.

Leaves Alternate, broadly oval with serrated margins, three lobes, hairy above.

Flowers Five petals, in clusters in the leaf axils.

Fruit/seed An ovoid capsule with hooked spines, densely hairy.

Flowering period
September–October.

Habitat
Monsoon areas of Dhofar, often near water; altitude from 150–900 m.

Distribution
Tropical Africa, Arabia, Pakistan to China.

Family Sterculiaceae (Cola or Sterculia Family)

Name

Melhania ovata (Cav.) Spreng

Description

A small, spreading, perennial shrub; up to 10 cm high and 30 cm wide.

Leaves Alternate, lance-shaped with slightly serrated margins, hairy.

Flowers Solitary in the leaf axils with long green bracts.

Fruit/seed Oval, five-lobed capsule.

Flowering period

Most of the year.

Habitat

Gravel plains and rocky crevices; altitude from 900–2,000 m.

Distribution

NE Africa, Arabia and east to Pakistan.

Family Zygophyllaceae (Creosote bush or Lignum-vitae Family)

Name

Tetraena simplex (L.) Beier & Thulin

Synonym
Zygophyllum simplex L.

Common names
abu rukaiba, arid, batbaak, harm, qarmal, ratreet

Description
Prostrate annual with red stems branching from the base to form a mat; 20–30 cm high.

Leaves Opposite, bright green/yellow, succulent, cylindrical, with entire margins.

Flowers Often solitary, bright yellow with five petals.

Fruit/seed Oval capsule with winged segments.

Flowering period
Most of the year.

Habitat
Common after rain in gravel and sandy areas; altitude up to 300 m.

Distribution
N and NE Africa, Arabia and east to India.

Uses
Leaves used to cure eye infections in UAE.

Family Zygophyllaceae (Creosote bush or Lignum-vitae Family)

Name

Tribulus terrestris L.

Common name
Puncture vine

Description
Woody, prostrate, hairy perennial with branching stems; up to 60 cm.

Leaves Opposite, compound with up to seven pairs of oblong leaflets.

Flowers Solitary, pale yellow with five free petals, on stalks arising from the nodes.

Fruit/seed Hairy, consisting of five segments, each with four distinctive spines.

Flowering period
Throughout the year.

Habitat
Shallow sands, saline areas, gravel and disturbed areas, common in urban areas; altitude up to 500 m.

Distribution
Tropics and sub-tropics.

Uses
Treatment of oral infections in N Africa.

Note
There are several species of *Tribulus* in Oman, generally identified by their fruit.

White

and cream flowers

Family Acanthaceae (Acanthus Family)

Name

Blepharis dhofarensis A.G.Mill.

Description

Large shrub or small spindly tree, much branched, older wood grey, young wood reddish; up to 5 m.

Leaves Opposite or whorled, elliptic with entire margins.

Flowers Emerge from a large cluster of spiny bracts.

Fruit/seed A flattened capsule. Seeds white or brownish with small hairs that swell when wet.

Flowering period

October–January.

Habitat

Dense thickets on the wet escarpment in Dhofar, often at the edges of cliffs; altitude from 500–900 m.

Distribution

Endemic to Dhofar and SE Yemen.

Uses

Seeds are good fodder for camels.

Note

Has a restricted distribution, therefore vulnerable.

Family	Amaranthaceae (Amaranthus, Celosias or Cockscomb Family)

Name

Aerva javanica (Burm.f.) Juss. ex Schult.

Common name
ra

Description
An erect, woody perennial; up to 1 m.

Leaves Alternate, lance shaped, greyish with entire margins, densely hairy.

Flowers Small, white/grey, covered in short, white hair, densely clustered on upright spikes.

Fruit/seed Shiny round, black or brown.

Flowering period
Most of the year.

Habitat
Common on disturbed ground; altitude up to 2,000 m.

Distribution
Dry areas of the Old World tropics and sub-tropics.

Uses
Used to stuff pillows in the Middle East.

Family Amaranthaceae (Amaranthus, Celosias or Cockscomb Family)

Name

Pupalia lappacea (L.) A.Juss.

Description

An erect annual or perennial covered with small hairs; up to 90 cm.

Leaves Opposite, variable in size and shape with entire margins and a pointed tip.

Flowers Woolly clusters on erect spikes with protruding white or pink/red hooked bracts.

Fruit/seed Ovoid capsule, shed with the flowers as a very sticky burr.

Flowering period

Most of the year.

Habitat

Along roadsides, disturbed ground and wadis; altitude up to 1,800 m.

Distribution

Widespread in the Old World tropics and sub-tropics.

Family Apiaceae (Carrot or Parsley Family)

Name

Ammi majus L.

Common names
Queen Anne's lace, nayniya

Description
An erect annual with grooved stems; up to 60 cm.

Leaves Alternate, deeply dissected lower leaves, upper leaves broadly serrated.

Flowers Clusters of white flowers form a terminal umbrella shape.

Fruit/seed Flat, oblong and bristly.

Flowering period
Most of the year.

Habitat
Common weed in wadis and plantations in Dhofar; altitude up to 2,200 m.

Distribution
NE Africa and Eurasia.

Note
Toxic. Food plant of the caterpillar of the swallowtail butterfly (*Papilio machaon*).

Family	Apocynaceae: Asclepiadoideae (Oleander or Milkweed Family)

Name

Rhazya stricta Decne.

Common name

harmal

Description

An erect perennial sub-shrub; up to 80 cm.

Leaves Alternate, lance shaped, yellowish with a clear mid-rib and entire margins.

Flowers Five free petals with pale blue buds, in clusters at the end of stems.

Fruit/seed Paired, yellowish/green with brown, winged seeds.

Flowering period

November–May.

Habitat

Common in sand and gravel areas and wadi beds throughout Oman; altitude up to 700 m.

Distribution

Deserts of Arabia, SW and central Asia.

Uses

Leaves have a variety of medicinal uses, both when crushed and mixed with boiling water or oil and when burned and the smoke inhaled.

Note

Avoided by grazing animals. Named after the eleventh century Arab physician Razi.

Family Apocynaceae: Asclepiadoideae (Oleander or Milkweed Family)

Name

Sarcostemma viminale (L.) R.Br.

Description
Perennial with scrambling stems forming a tangled clump; up to 1 m.

Leaves Minute scales, appears leafless.

Flowers Five cream/white fleshy lobes.

Fruit/seed Paired with flattened seeds.

Flowering period
Most of the year.

Habitat
Common on the drier escarpment in Dhofar and the eastern Hajar mountains in northern Oman; altitude up to 1,300 m.

Distribution
S Africa, SW Arabia, Pakistan, Nepal, India and Australia.

Uses
New growth after rain provides a popular and refreshing food supplement.

Family Asphodelaceae (Aloe Family)

Name

Asphodelus tenuifolius Cav.

Description

An erect annual with branched stems arising from small bulbs; 15–70 cm.

Leaves Numerous, arising from the base, linear, hollow.

Flowers Six white petals with a dark red stripe down the centre, on erect terminal spikes.

Fruit/seed Small brown capsules with many seeds.

Flowering period

December–May.

Habitat

Widespread in sand and gravel areas in northern Oman; altitude up to 1,650 m.

Distribution

Canary Islands, the Mediterranean, E Africa and Arabia.

Uses

The whole plant can be boiled in water and used as a laxative or diuretic.

Family Asteraceae (Daisy or Sunflower Family)

Name

Bidens pilosa L.

Common name
Blackjack

Description
Erect annual, branching above; up to 1 m.

Leaves Trifoliate with serrated margins.

Flowers White petals with yellow centres, often remain closed.

Fruit/seed Black, barbed.

Flowering period
Most of the year.

Habitat
Open sunny areas, often near water and as a weed of cultivation; altitude up to 2,200 m.

Distribution
Pan-tropical weed.

Uses
Many medicinal uses throughout the tropics.

Family Asteraceae (Daisy or Sunflower Family)

Name

Eclipta prostrata L.

Common name
False daisy, masadate

Description
A small prostrate herb with much branched stems; spreading up to 1 m.

Leaves Opposite, elliptical with entire or finely serrated margins, rough, hairy.

Flowers Solitary or paired flower heads arise from the leaf nodes.

Fruit/seed Black or brown, triangular, winged achene.

Flowering period
December–April.

Habitat
Disturbed areas around plantations and permanently damp places; altitude up to 900 m.

Distribution
Native to tropical America, now a pan-tropical weed.

Uses
In India it is used against nematodes.

Family	Asteraceae (Daisy or Sunflower Family)

Name

Parthenium hysterophorus L.

Common name
Feverfew

Description
A fast growing annual with a strong scent that repels insects; up to 1.5 m.

Leaves Opposite, pale green, deeply dissected, with soft hairs.

Flowers Small, white, with five distinct corners, in terminal or axillary clusters.

Fruit/seed Four or five black, wedge-shaped seeds from each flower.

Flowering period
All year.

Habitat
Sandy areas of wadis and as a weed of cultivation; altitude up to 900 m.

Distribution
A pan-tropical weed, originally native to C and tropical S America.

Note
This plant is strongly allergenic to humans and animals. It releases chemicals that inhibit the growth of other plants. It has a deep tap root that is difficult to remove.

Family Asteraceae (Daisy or Sunflower Family)

Name

Tridax procumbens L.

Common name
Coat buttons

Description
A sprawling, hairy perennial with milky sap; up to 40 cm.

Leaves Opposite, hairy, with irregular lobes and coarsely serrated margins.

Flowers White flowers with yellow centres, borne on long erect hairy stalks.

Fruit/seed Black, hairy achene with bristles on top.

Flowering period
Most of the year.

Habitat
Among rocks in moist wadis in Dhofar; altitude up to 800 m.

Distribution
Pan-tropical weed, originally native to C and tropical S America.

Family Boraginaceae (Borage or Forget-me-not Family)

Name

Echiochilon persicum (Burm.f.) I.M Johnst.

Common name
Stone-seed

Description
Small woody perennial with stems branched from the base; up to 30 cm.

Leaves Alternate, grey/green, lance-shaped, with dense white hair and a pointed tip.

Flowers Solitary arising in the leaf nodes with no stalk.

Fruit/seed Small brown nutlets.

Flowering period
January–May.

Habitat
Sand and gravel areas; altitude up to 1,500 m.

Distribution
E Africa, Arabia, Iran and Pakistan.

Family Boraginaceae (Borage or Forget-me-not Family)

Name
Heliotropium bacciferum Forssk.

Common name
hahhfa

Description
A dense, woody perennial covered in stiff white hairs; up to 1 m.

Leaves Alternate, narrow, lance shaped with an acute tip and wavy margins.

Flowers Dense, double rows on one side of a curved stalk.

Fruit/seed Round, white berry, with two pairs of nutlets.

Flowering period
Most of the year.

Habitat
Common in sand and gravel areas; altitude up to 200 m.

Distribution
N and W Africa, Arabia and east to India.

Uses
Fruit is edible.

Similar species
There are several species of *Heliotropium* in Oman.

Family Boraginaceae (Borage or Forget-me-not Family)

Name

Heliotropium longiflorum (A. DC.) Jaub. & Spach

Description

An erect herb with hairy stems; up to 40 cm.

Leaves Alternate, oval with an acute tip and wavy margins, prominent nerves with small hairs.

Flowers Small, grouped along one side of a forked, curved stalk.

Fruit/seed Round, consisting of four nutlets.

Flowering period

Most of the year.

Habitat

Damp disturbed or open ground in Dhofar; altitude up to 900 m.

Distribution

Tropical Africa and Arabia.

Note

The name is from the Greek meaning 'turning to the sun'.

Family Boraginaceae (Borage or Forget-me-not Family)

Name

Trichodesma hildebrandtii Gürke

Description

A perennial herb covered in white bristles; up to 50 cm.

Leaves Opposite, oblong with an acute tip and entire margins.

Flowers White with yellow and brown blotches in the centre of the five fused petals, face downwards.

Fruit/seed Four nutlets, with spiny margins, enclosed in an enlarged calyx.

Flowering period

Most of the year, if near water.

Habitat

Coastal and dry areas of the Dhofar escarpment; altitude up to 500 m.

Distribution

Kenya, Ethiopia, Somalia, Yemen and Oman.

Uses

Traditionally used to treat mastitis in animals and women.

Note

There are several species of Trichodesma in Oman. Some species contain alkaloids which are associated with causing liver cancer. The name comes from the Greek *trichos* meaning 'hair' and *desma* meaning 'bundle' and refers to the way the anthers are held together by two rows of hairs.

Family	Brassicaceae (Cabbage or Mustard Family)

Name

Morettia parviflora Boiss.

Common name
hafra

Description
Woody based perennial covered in white hairs; up to 40 cm.

Leaves Grey/green, broadly oval with entire margins.

Flowers Small, white with yellow centres, borne terminally.

Fruit/seed Hairy, curved away from the stem.

Flowering period
November–May.

Habitat
Common in sand and gravel areas, wadis, desert and disturbed ground; altitude up to 1,800 m.

Distribution
Egypt, Sinai, Palestine, Arabia and Jordan.

Family Capparaceae (Caper Family)

Name

Capparis cartilaginea Decne.

Common names
Caper plant, lezaf

Description
Climbing or sprawling shrub, spiny, often found hanging from rocky cliffs.

Leaves Alternate, simple, with entire margins and a rather leathery texture, often with paired hooked spines at the base of the stalk.

Flowers Showy white, with many protruding stamens.

Fruit/seed Oval, green turning deep red as it matures.

Flowering period
Most of the year.

Habitat
Common on rocky cliffs; altitude up to 2,000 m.

Distribution
Tropical Africa to India.

Uses
The fruit is high in vitamin C. Crushed leaves have a variety of medicinal uses.

Note
Attracts caper white (*Anaphaeis aurota*) and desert orange tip (*Colotis liagora*) butterflies.

Family Caryophyllaceae (Pink or Carnation Family)

Name

Cometes surattensis L.

Description

A low growing densely hairy, cushion-shaped annual; up to 20 cm.

Leaves Oval with entire margins, very hairy.

Flowers Form a mass of loose irregular sprays covered in long white hairy bracts.

Fruit/seed Cylindrical capsule.

Flowering period

Most of the year, especially after rain.

Habitat

Gravel wadis, mountain slopes and disturbed areas; altitude up to 900 m.

Distribution

Egypt, Sinai, Iraq, Iran, Arabia and Pakistan.

Family Caryophyllaceae (Pink or Carnation Family)

Name

Gypsophila montana Balf.f.

Description
A cushion shaped, erect perennial with a woody base; up to 80 cm.

Leaves Opposite, oval, usually forming a basal rosette.

Flowers Many tiny flowers, forming sprays at the end of leafless stalks.

Fruit/seed A capsule with four valves. Seeds rough with a shallow groove at the back.

Flowering period
Most of the year.

Habitat
Open rocky slopes and gravel wadis in Dhofar; altitude from 100–1,200 m.

Distribution
Southern Arabia and Somalia.

Family Caryophyllaceae (Pink or Carnation Family)

Name

Spergula fallax (Lowe) E.H.L.Krause

Description

Annual with slender branching stems; up to 25 cm.

Leaves Whorled, linear with entire margins.

Flowers Green sepals surround white petals with pointed tips.

Fruit/seed Capsule with minute, winged seeds.

Flowering period

February–April.

Habitat

In sandy soil, wadi beds and as a weed of cultivation in northern Oman; altitude up to 1,200 m.

Distribution

Canary Islands, Mediterranean, NE Africa, Middle East, Arabia, and SW Asia.

Family Convolvulaceae (Morning-glory or Bindweed Family)

Name

Convolvulus cephalopodus Boiss.

Description

Soft woolly, spreading perennial; up to 50 cm.

Leaves Alternate, grey, covered in white hairs, with wavy margins, crowded towards the end of stems.

Flowers Pale mauve, trumpet shaped with fused, slightly wrinkled petals that arise from the leaf axils.

Fruit/seed A capsule with hairy black seeds.

Flowering period

January–May.

Habitat

Common on coastal sand dunes, also found on low gravel plains; altitude up to 100 m.

Distribution

UAE, Iran and Pakistan.

Note

There are 17 species of *Convolvulus* in Oman, all with similar flowers, some have a blue or pink tinge.

Family Convolvulaceae (Morning-glory or Bindweed Family)

Name

Convolvulus virgatus Boiss.

Common name
hibab risha

Description
A dense spreading shrub with spiny branches, almost leafless; up to 60 cm.

Leaves Alternate, small, lance shaped with entire margins.

Flowers May be clustered, axillary, close in the afternoon.

Fruit/seed Smooth capsule. Seeds dark brown.

Flowering period
Most of the year.

Habitat
Rocky slopes, sand and gravel areas; altitude up to 600 m.

Distribution
UAE and S Iran.

Family Convolvulaceae (Morning-glory or Bindweed Family)

Name

Cuscuta planiflora Ten.

Common names
Dodder, cow lace

Description
A parasite without chlorophyll or leaves, thread-like yellowish/green stems twine around various low-growing host plants.

Leaves Absent.

Flowers Small, growing in clusters along the stems.

Fruit/seed A small capsule with ovoid seeds.

Flowering period
Most of the year, if near water.

Habitat
Climbing over woody plants in sand and gravel areas and as a weed of cultivation; altitude up to 150 m.

Distribution
Canary Islands, the Mediterranean, Egypt, Arabia and W Asia.

Note
Grows from a seed but the roots disappear as soon as it finds a host plant.

Family Convolvulaceae (Morning-glory or Bindweed Family)

Name

Hildebrandtia africana Vatke

Description

A small very dense shrub with dark grey to black bark; up to 1 m.

Leaves Alternate, oval with a rounded tip and entire margins, covered in silky hair.

Flowers Four white petals surrounded by two large pale green sepals which turn brown on maturity.

Fruit/seed A round capsule with four seeds.

Flowering period

January–April and September–November.

Habitat

Rocky slopes, gravel wadis and the escarpment woodlands of Dhofar; altitude up to 1,400 m.

Distribution

Tropical NE Africa and southern Arabia.

Note

Named after J.M. Hildebrandt (1847–1881) who made collections in north and east Africa.

Family Convolvulaceae (Morning-glory or Bindweed Family)

Name

Seddera glomerata (Balf. f)
O. Schwartz

Description
Perennial dwarf shrub; up to 30 cm.

Leaves Silver/grey densely covered with small hairs; oval with an acute tip and entire margins.

Flowers Spherical, compact heads with white to pink flowers, terminal.

Fruit/seed A hairy capsule.

Flowering period
September–October.

Habitat
In dry sand and gravel areas; altitude up to 1,600 m.

Distribution
Endemic to Yemen and southern Oman.

Family Euphorbiaceae (Spurge Family)

Name

Euphorbia orbiculifolia S.Carter

Description
Succulent, perennial shrub, stems covered in leaf scars; up to 20 cm.

Leaves Opposite, clustered at the end of branches, round with entire margins.

Flowers Creamy/white with a pink base, five lobes and protruding stamens.

Fruit/seed A lobed capsule that hangs down.

Flowering period
September–October.

Habitat
On the Dhofar escarpment; altitude from 200–1,400 m.

Distribution
Endemic to Dhofar and SE Yemen.

Note
Due to its restricted distribution this species is vulnerable.

Family Euphorbiaceae (Spurge Family)

Name

Jatropha dhofarica Radcl.-Sm.

Description

An erect shrub, stems have clear or red latex, older stems with leaf scars; up to 3 m.

Leaves Alternate, bright green, palmately lobed with serrated or wavy margins.

Flowers Small, cream with a pink centre, in clusters at the end of branches.

Fruit/seed Oval capsule containing four large, brown seeds.

Flowering period

Most of the year.

Habitat

Common, especially in open woodland clearings on the wet escarpment of Dhofar; altitude up to 1,200 m.

Distribution

Endemic to Dhofar and SE Yemen.

Uses

The sap is used as an antiseptic and to provide protection to cut surfaces. The seeds are edible with a pleasant nutty flavour.

Family	Euphorbiaceae (Spurge Family)

Name

Ricinus communis L.

Common names
Castor oil plant, arash

Description
Erect perennial with woody base and a single hollow stem; up to 2 m.

Leaves Alternate, glossy green, palmately lobed with serrated margins.

Flowers Red female above and white male flowers below on a terminal spike.

Fruit/seed Round, white capsules with dark spines. Seeds mottled brown.

Flowering period
Most of the year, if near water.

Habitat
Disturbed ground and wadis; altitude up to 800 m.

Distribution
Native to tropical NE Africa and the Middle East, now pan-tropical.

Uses
Cultivated commercially for its oil (from the seeds), which has numerous medicinal and industrial purposes.

Note
The seeds contain the poison Ricin which is water soluble and not present in the oil.

Family Iridaceae (Iris Family)

Name

Gladiolus candidus (Rendle) Goldblatt

Synonym
G. ukambanensis (Baker) Marais

Description
An erect annual, rising from a small bulb; up to 45 cm.

Leaves Alternate, long, thin, grass-like.

Flowers Long, funnel shaped, opening into six free petals.

Fruit/seed A capsule with reddish seeds.

Flowering period
August–October.

Habitat
Common on the escarpment mountains of Dhofar; altitude from 200–1,000 m.

Distribution
Kenya, Tanzania, Ethiopia and Somalia.

Uses
Bulbs are a valued food.

Note
The name is derived from the Latin *gladiolus* meaning 'sword'.

Family Lamiaceae (Mint Family)

Name

Leucas inflata Benth.

Common name
qotnya

Description
Annual or perennial with square erect stems branching from the base, covered in stiff white hair; up to 80 cm.

Leaves Opposite, oval with wavy margins.

Flowers Whorled clusters on a terminal spike.

Fruit/seed Ovoid nutlets.

Flowering period
December–April.

Habitat
Wadis, gravel plains and hillsides; altitude up to 1,500 m.

Distribution
Egypt, Sudan, Ethiopia and Arabia.

Family Lamiaceae (Mint Family)

Name

Leucas urticifolia R.Br.

Description
An erect annual, usually hairy; up to 80 cm.

Leaves Opposite, bright green, oval with serrated edges.

Flowers In dense whorls in the leaf nodes.

Fruit/seed Small brown nutlets.

Flowering period
August–October.

Habitat
Common on the wet escarpment of Dhofar; altitude up to 600 m.

Distribution
E Africa, southern Arabia and east to India.

Uses
Used to treat fever in India.

Family Lythraceae (Loosetrife, Pomegranate, Henna, Water Chestnut or Crepe-myrtle Family)

Name

Lawsonia inermis L.

Common names
Henna, hinna

Description
An untidy shrub, older branches may be spiny; up to 3 m.

Leaves Opposite, elliptical with entire margins.

Flowers White/cream in terminal clusters, with a strong scent.

Fruit/seed Berry-like, red, turning dark brown as they mature.

Flowering period
All year.

Habitat
Naturalised in wadis and sandy and disturbed areas; altitude up to 2,000 m.

Distribution
Widespread in Arabia, Africa and Asia, introduced to the New World tropics.

Uses
Crushed leaves mixed with lime juice produce a dye for skin and hair. Oil from the flowers is used in perfumes in Arabia.

Note
Named after Dr Isaac Lawson who financed the publication of Linnaeus's 'Systema Naturae'.

Family Malvaceae (Mallow, Cotton or Hollyhock Family)

Name

Abelmoschus manihot (L.) Medik.

Common names
Okra, semerhot

Description
Erect annual herb with reddish stems; up to 1 m.

Leaves Alternate, palmately lobed with slightly serrated margins.

Flowers Five overlapping petals with a red centre, borne in leaf axils.

Fruit/seed Ovoid capsule, with a narrow tip, covered in hair.

Flowering period
September–November.

Habitat
Common in the tall grasslands of the escarpment in Dhofar; altitude up to 900 m.

Distribution
Originally from the Old World tropics, now pan-tropical. Widely cultivated as a vegetable.

Uses
The seeds and young pods are eaten raw and cooked when older.

Note
A. esculentus is similar with larger flowers and leaves and a beaked fruit.

Family Malvaceae (Mallow, Cotton or Hollyhock Family)

Name

Malva parviflora L.

Common names
Cheeseweed mallow, shuwayb al hamam

Description
Annual herb, prostrate, branched, spreading stems; up to 25 cm high.

Leaves Alternate, overall round, shallowly lobed with indented base, margins serrated.

Flowers White with a tinge of pink, may be in clusters of up to 6.

Fruit/seed Round with 7–11 flat sections.

Flowering period
February–April.

Habitat
Disturbed and cultivated areas; altitude up to 2,200 m.

Distribution
Native to Europe and Asia, now a cosmopolitan weed.

Family Malvaceae (Mallow, Cotton or Hollyhock Family)

Name

Pavonia schweinfurthii Ulbr.

Description An erect perennial with branched stems; up to 65 cm.

Leaves Alternate, simple, narrowly oval with very slightly serrated margins that curl inwards.

Flowers Single, axillary, cream with five petals.

Fruit/seed Hairy sections enclosed in an epicalyx.

Flowering period
Most of the year.

Habitat
On seaward facing wadis and gravel slopes in Dhofar; altitude up to 200 m.

Distribution
NE Africa and southern Oman; not recorded in other parts of Arabia.

Family Oleaceae (Olive Family)

Name

Jasminum grandiflorum L.

Common names

Jasmine, yasmin razqy

Description

A weak sprawling and climbing shrub trailing over other plants; stems up to 3 m long.

Leaves Opposite, five-foliate with the terminal leaf larger than the two lateral ones, margins entire.

Flowers Sweet smelling, in small groups either terminally or in the leaf axils.

Fruit/seed Fleshy, divided into two lobes, each with one seed.

Flowering period

Most of the year.

Habitat

The wet escarpment of Dhofar; altitude from 150–900 m.

Distribution

E Africa, Sudan, and southern Arabia.

Uses

Flowers are widely used as a perfume, the leaves have disinfectant medicinal properties.

Family Plumbaginaceae (Leadwort, Sea-lavender or Thrift Family)

Name

Plumbago zeylanica L.

Description

A scrambling perennial herb, stems smooth near the base, covered in small glands higher up near the flowers; up to 2 m.

Leaves Alternate, simple, elliptic with entire margins.

Flowers A tubular corolla opening into five petals cupped in a long narrow green calyx covered in sticky glands.

Fruit/seed Small, cylindrical with a pointed apex and persistent calyx.

Flowering period

Most of the year.

Habitat

The wet escarpment in Dhofar; altitude from 200–1,100 m.

Distribution

Native to the Old World Tropics, now pan-tropical.

Note

Has antimicrobial and antifungal properties.

Family Polygonaceae (Dock, Buckwheat or Rhubarb Family)

Name

Calligonum comosum L'Hér.

Common names
abal, arta, dhakar

Description
An almost leafless shrub with pale branches and swollen nodes; up to 2 m.

Leaves Minute, soon falling.

Flowers Small, white with a green median stripe, sometimes in groups, arising from the leaf axils.

Fruit/seed Showy, with four pairs of longitudinal wings covered in stiff bright red bristles.

Flowering period
January–April.

Habitat
Sandy areas and dunes in northern Oman, dominant in areas with deep sand; altitude up to 200 m.

Distribution
Common in desert areas from Egypt to Pakistan.

Uses
Nearly smokeless and therefore valued as firewood. Used to cure leather. The fruit is edible.

Family Resedaceae (Mignonette or Weld Family)

Name

Reseda aucheri Boiss.

Common name
zinban

Description
Erect, branching leafy annual or perennial; up to 60 cm.

Leaves Alternate, slightly fleshy, oval, with wavy margins.

Flowers Small white/cream in dense conical spikes.

Fruit/seed Ovoid capsule with three teeth at the top. Seeds minute and shiny.

Flowering period
January–May.

Habitat
Sandy or dry rocky areas in northern Oman; altitude up to 1,300 m.

Distribution
Turkey, Syria, Palestine, Arabia, W and C Asia.

Note
Attracts the desert white butterfly (*Pontia glauconome*).

Family Rubiaceae (Coffee or Madder Family)

Name

Gaillonia aucheri (Guill.) Jaub. & Spach

Common name
khurman

Description
A woody shrub with white bark, stiff branches intertwined with smaller twigs tapering into spines, the whole plant has an unpleasant smell; up to 1.5 m.

Leaves Whorled, cylindrical, succulent, small.

Flowers Small, trumpet shaped, clustered in the leaf axils, obscured by a mass of white hairs, giving the plant a woolly appearance.

Fruit/seed Splits into two when ripe.

Flowering period
January–May.

Habitat
Gravel plains and wadis; altitude up to 1,700 m.

Distribution
Arabia to SE Asia.

Family Scrophulariaceae (Foxglove or Figwort Family)

Name

Orobanche dhofarensis Foley

Description
Erect parasite with a pinkish hairy stem; up to 40 cm.

Leaves Absent.

Flowers Funnel shaped, spaced alternately on the stem.

Fruit/seed A capsule containing many dust-like seeds.

Flowering period
August–October.

Habitat
Wet areas of the Dhofar escarpment; altitude from 200–1,100 m.

Distribution
Endemic to Dhofar.

Family	Solanaceae (Potato Family)

Name

Datura metel L.

Common names
Devil's trumpet, maranhah

Description
A fast growing annual; up to 2 m.

Leaves Alternate, oval with broadly serrated margins.

Flowers Solitary, white, trumpet shaped with five fused petals.

Fruit/seed Green, rounded and covered with short blunted spines. Seeds yellow.

Flowering period
Most of the year where there is sufficient moisture.

Habitat
Moist, sunny areas where the soil contains clay, often as a weed in cultivated and disturbed areas; altitude up to 2,200 m.

Distribution
Native to the Americas, now pan-tropical.

Note
Very toxic in overdose but used in moderation as a sedative, the seeds in particular are narcotic.

Family Solanaceae (Potato Family)

Name

Lycium shawii Roem. & Schult.

Common name

qasad

Description

Spiny, intricately branched, deciduous shrub; up to 2 m.

Leaves Alternate or in clusters, oblong with entire margins.

Flowers Small, funnel-shaped, white or mauve, in the leaf axils.

Fruit/seed Red berries with many yellow seeds.

Flowering period

January–April.

Habitat

Common throughout Oman; altitude up to 1,000 m.

Distribution

Egypt, Sudan, Palestine and Arabia.

Uses

Good fuel, producing little smoke. Historically *Lycium* species were used in fumigation for a variety of complaints.

Note

Lycium is from the Greek meaning 'spiny'.

Family Solanaceae (Potato Family)

Name

Physalis minima L.

Description
An erect branched annual; up to 40 cm.

Leaves Alternate, elliptical, slightly hairy with entire or irregularly serrated margins.

Flowers Cream with five fused petals, borne on a thin stalk from the leaf axil.

Fruit/seed Berry enclosed in a papery 'lantern' that hangs down.

Flowering period
November–May.

Habitat
A weed at the edges of cultivated land; altitude up to 1,250 m.

Distribution
Arabia, SE Asia and Australia.

Family Solanaceae (Potato Family)

Name

Solanum nigrum L.

Common names
Black nightshade, majaj, shajarat al gharub

Description An erect annual with slender branched stems; up to 60 cm.

Leaves Alternate, oval with entire or slightly wavy margins, sometimes hairy.

Flowers Five triangular petals, often curved backwards from a yellow centre.

Fruit/seed A berry turning from green to red and finally black as it ripens.

Flowering period
Most of the year.

Habitat
Near water and around human habitation; altitude up to 2,300 m.

Distribution
Native to Europe, now widely naturalised.

Uses
Crushed black berries used as a skin dye.

Note
Toxic to livestock.

Family Sparrmanniaceae (Jute, Sparrmannia or Grewia Family)

Name

Grewia erythraea Schweinf.

Common name
sharham

Description
A prostrate dense, woody bush with grey bark; up to 2 m.

Leaves Alternate, oval, with serrated margins.

Flowers Five free white petals bent back revealing many stamens, fragrant.

Fruit/seed Green, turning orange as they mature.

Flowering period
Most of the year.

Habitat
The northern mountains and on the dry plateau in Dhofar; altitude up to 1,850 m.

Distribution
NE Africa to Pakistan and India.

Note
There are four *Grewia* species in Oman, two of which have yellow flowers and two white.

Family Violaceae (Violet Family)

Name

Hybanthus durus (Bak.) O.Schwartz

Description
A small dense shrub, young branches have white bark; up to 1 m.

Leaves Alternate, oval with rounded tips and entire margins.

Flowers Four small upper petals are violet, the lower, relatively large fifth petal is white with a pale mauve stripe, spoon shaped.

Fruit/seed A small smooth capsule.

Flowering period
September–November.

Habitat
On the Dhofar escarpment; altitude from 300–1,300 m.

Distribution
Somalia, Ethiopia and southern Arabia.

Family Vitaceae (Grape vine or Virginia creeper Family)

Name

Cissus quadrangularis L.

Common name
ataerey

Description
A perennial climber with square, thick, succulent stems, constricted at the nodes, climbs over other plants; up to 5 m.

Leaves Alternate, oval with three lobes and slightly serrated margins, simple tendrils emerge opposite the leaves.

Flowers Cream or pale yellow, may have red tips, arise opposite the leaves and hang down in clusters.

Fruit/seed Berry-like, bright red when ripe.

Flowering period
Most of the year.

Habitat
Common on rocky outcrops in Dhofar; altitude from 100–1,000 m.

Distribution
Tropical Africa, southern Arabia and India.

Uses
The pounded plant is used to cure surface wounds and mastitis in goats.

Note
Cyphostemma ternatum is similar, but can be distinguished by its trifoliate leaves.

Family Zygophyllaceae (Mallow, Cotton or Hollyhock Family)

Name

Tetraena qatarensis (Hadidi) Beier & Thulin

Synonym
Zygophyllum qatarense Hadidi

Description
A dense, cushion-forming bright green perennial; up to 80 cm.

Leaves Opposite, shiny green or yellowish, cylindrical, fleshy.

Flowers Inconspicuous, five white petals.

Fruit/seed Oblong capsule with a square top, yellow to brown.

Flowering period
January–March.

Habitat
Very common on dry sand and gravel plains throughout the country; altitude up to 500 m.

Distribution
Endemic to Arabia.

Green

and brown flowers

Family Acanthaceae (Acanthus Family)

Name

Ecbolium viride (Forssk.) Alston

Common name
thawmah

Description
A tall, bushy herb; up to 1 m.

Leaves Opposite, dark green, broadly lance-shaped with entire margins.

Flowers Greenish/white, borne between dense bright green bracts on an upright terminal spike.

Fruit/seed Capsule. Seeds large, green with a lumpy surface.

Flowering period
Most of the year.

Habitat
Disturbed areas along roads or wadis; altitude up to 1,500 m.

Distribution
Tropical Africa, southern Arabia and India.

Family Amaranthaceae (Amaranthus, Celosias or Cockscomb Family)

Name

Alternanthera pungens Kunth

Description
Prostrate, hairy perennial forming mats along the ground; up to 20 cm high, spreading to 1 m.

Leaves Opposite, oval with entire margins.

Flowers Inconspicuous, borne in the leaf axils and surrounded by spiny hairs.

Fruit/seed A round capsule.

Flowering period
Most of the year.

Habitat
Disturbed areas along roadsides and compacted sand or gravel in Dhofar; altitude up to 300 m.

Distribution
Native to tropical America, now pan-tropical.

Family Amaranthaceae (Amaranthus, Celosias or Cockscomb Family)

Name

Amaranthus graecizans L.

Common names
dadh, quttaif, shegra tal santeen, sindar

Description
Erect annual, stems may be reddish; up to 70 cm.

Leaves Alternate with long stalks, oval with entire margins.

Flowers Green/red clusters borne on branching spikes.

Fruit/seed Green/red, almost round. Seeds black and shiny.

Flowering period
Most of the year.

Habitat
Very common, found on disturbed ground and along roadsides; altitude up to 1,500 m.

Distribution
Pan-tropical weed.

Note
A. viridis is similar with flowers borne on terminal spikes. *Amaranthus* comes from the Greek meaning 'not fading' and refers to the flowers of some species which appear to be everlasting.

Family Apocynaceae: Asclepiadoideae (Oleander or Milkweed Family)

Name

Gomphocarpus fruticosus (L.) W.T.Aiton

Description

An erect shrub with white latex; up to 1 m.

Leaves Opposite, oblong with an acute tip and entire margins.

Flowers Pendulous clusters, with 4–7 individual flowers.

Fruit/seed Single or paired, club-shaped with soft bristles.

Flowering period

June–November.

Habitat

The escarpment woodlands and along roadsides in Dhofar; altitude from 600–1,100 m.

Distribution

Africa, the Mediterranean and Arabia.

Uses

Latex used to remove hair from hides.

Note

Toxic to livestock.

Family Apocynaceae: Asclepiadoideae (Oleander or Milkweed Family)

Name

Pergularia tomentosa L.

Common names
ghalaqah, shajarat al julud

Description
A twining perennial, exudes white latex when cut; climbs up to 2 m.

Leaves Opposite, heart-shaped with entire margins.

Flowers Five petals with white hair, hang in groups of 2–3 flowers.

Fruit/seed Paired, club-shaped, very pale green with soft bristles.

Flowering period
Most of the year.

Habitat
Sand and gravel plains, wadis and mountain slopes; altitude up to 1,500 m.

Distribution
Tropical N Africa, Arabia and east to Pakistan.

Uses
Latex used to treat skin infections and to remove hair from hides.

Note
Attracts the plain tiger butterfly *(Danaus chrysippus)*.

Family	Caryophyllaceae (Pink or Carnation Family)

Name

Herniaria maskatensis Bornm.

Description Small perennial herb, dull green, lies flat on the ground; up to 7 cm.

Leaves Opposite, broadly round with dense glands.

Flowers Minute in clusters along the stem.

Fruit/seed One seeded round capsule.

Flowering period
February–April.

Habitat Rocky and gravel areas; altitude up to 1,300 m.

Distribution
Endemic to Oman and SE Yemen.

Family Chenopodiaceae (Goosefoot, Beetroot or Spinach Family)

Name

Anabasis setifera Moq.

Description
Much branched perennial with fleshy young shoots, square branches with a conspicuous groove along the centre of each side; up to 60 cm.

Leaves Opposite, succulent, truncated, club-like.

Flowers Small brownish, borne in clusters of 3–7 in the leaf nodes.

Fruit/seed Pale yellow to red with five unequal wings.

Flowering period
October–March.

Habitat
Common in sand and gravel areas in northern and central Oman; altitude up to 250 m.

Distribution
Egypt, Arabia, SW Asia.

Note
One of the first plants to colonise disturbed coastal sites.

Family Chenopodiaceae (Goosefoot, Beetroot or Spinach Family)

Name

Arthrocnemum macrostachyum
(Moric.) K. Koch

Common name
hamadh

Description
A dense succulent perennial forming large clump-like mounds. Old stems woody, young ones succulent and segmented, they may be yellow, reddish or green; up to 2 m.

Leaves Reduced to small lobes, inconspicuous.

Flowers Minute three flowered clusters in the leaf axils, enclosed by bracts.

Fruit/seed Spongy with very small, black seeds.

Flowering period
November–February.

Habitat
Common in sandy coastal areas, mud flats and inland salt marshes; altitude sea level.

Distribution
The Mediterranean, Arabia and east to Pakistan and India.

Family Chenopodiaceae (Goosefoot, Beetroot or Spinach Family)

Name

Atriplex farinosa Forssk.

Common name
afar

Description
An erect bushy perennial with grey stems, whole plant covered in whitish powder; up to 1.5 m.

Leaves Alternate, oblong, smooth, pale bluish/grey, succulent, with white powder on both sides.

Flowers Small pale green with a pink tinge at the base, clustered in grey/red bracts.

Fruit/seed Enclosed in flat circular bracteoles.

Flowering period
October–February.

Habitat
Common in sandy coastal areas; altitude sea level.

Distribution
Tropical NE Africa, Jordan, Egypt and Arabia.

Family Chenopodiaceae (Goosefoot, Beetroot or Spinach Family)

Name

Chenopodium murale L.

Common names
Nettle-leaved goosefoot, al dhorbaih, al zorbaih

Description
An erect annual; up to 60 cm.

Leaves Alternate, roughly arrow shaped with irregularly serrated margins.

Flowers Tiny green, in loose or compact clusters either terminally or in leaf axils.

Fruit/seed Round with small black, shiny seeds.

Flowering period
Throughout the year.

Habitat
Common in disturbed and irrigated areas; altitude up to 2,300 m.

Distribution
Native to the Old World tropics, now cosmopolitan.

Uses
Young leaves are used in salads.

Family Euphorbiaceae (Spurge Family)

Name

Acalypha indica L.

Description
An erect annual; up to 75 cm.

Leaves Alternate, oval with an acute tip and serrated margins, on long stalks.

Flowers Small, green in groups of 3–5, on spikes borne in leaf axils or terminally.

Fruit/seed Hairy with three lobes. Seeds small, brown.

Flowering period
August–February.

Habitat
A weed in damp, cultivated areas and waste ground; altitude up to 900 m.

Distribution
Widespread in the Old World tropics, introduced to the New World.

Uses
Has a pleasant lemony flavour, used in cooking. A variety of medicinal uses are reported.

Note
The name *Acalypha* is from the Greek meaning 'not nice to touch' and was used by Linnaeus as the leaf resembles a stinging nettle.

Family Euphorbiaceae (Spurge Family)

Name

Chrozophora oblongifolia (Del.) A.Juss.

Common names
tannum, tinib

Description
A small woody perennial covered in stiff hairs; up to 50 cm.

Leaves Alternate, grey/green, oblong with irregular margins and pointed tips, hairy, stalks and leaf are of equal length.

Flowers Inconspicuous, yellowish/green, borne in the leaf axils or terminally.

Fruit/seed Hairy capsule with three lobes. Seeds dark brown.

Flowering period
October–May.

Habitat
Common in disturbed areas; altitude up to 1,550 m.

Distribution
NE Africa, Sinai, Arabia and east to Pakistan.

Uses
The blue/black juice intensifies in colour with exposure to air and was formerly used as ink in Arabia.

Family Euphorbiaceae (Spurge Family)

Name

Dalechampia scandens L.

Description
A woody, low-growing shrub; the hairy stems may trail up to 3 m.

Leaves Alternate, oblong, palmately lobed with serrated edges.

Flowers Cupped in two pale green bracts with serrated edges and dark green veins.

Fruit/seed Covered in short white hairs. Seed orange.

Flowering period
September–April.

Habitat
Sand and gravel plains and in wadis; altitude up to 850 m.

Distribution
Native to Panama, now naturalised in Cape Verde, tropical Africa, Arabia and east to India.

Note
If touched, hair may cause skin irritation.

Family Euphorbiaceae (Spurge Family)

Name

Euphorbia balsamifera Ait.

Description

Dome-shaped, deciduous bush, with copious white sap, bark grey and gnarled bearing leaf scars; up to 1 m high and 1.5 m wide.

Leaves Alternate, slightly fleshy with entire margins, in terminal rosettes.
Flowers Borne in the centre of the leaves, female yellow, male reduced to yellow stamens.
Fruit/seed A round capsule with brown seeds.

Flowering period

August–November.

Habitat

On the Dhofar escarpment, in the transition zone where the monsoon fog is less pronounced; altitude up to 1,200 m.

Distribution

Southern Arabia and Somalia.

Uses

Sap used as an adhesive, to extract body hair, and as chewing gum.

Family Euphorbiaceae (Spurge Family)

Name

Euphorbia hadramautica Baker

Description

Small succulent herb with cylindrical enlarged stems bearing leaf scars, exudes white latex when cut; up to 25 cm.

Leaves Linear with wavy margins, crowded at the top of the stem.

Flowers Small, green, in clusters in leaf axils, with round flattened bracts.

Fruit/seed A round hairy capsule with three lobes. Seeds brown with four angles.

Flowering period

June–November.

Habitat

Common on the drier parts of Dhofar; altitude up to 1,100 m.

Distribution

N Kenya, Somalia, Ethiopia and Yemen.

Note

Originally collected in Hadramaut (southern Arabia) by William Lunt. Toxic.

Family Euphorbiaceae (Spurge Family)

Name

Euphorbia larica Boiss.

Common name
isbaq

Description
A deciduous, dense erect shrub with yellow/green, branching stems which exude thick white latex if broken; up to 1 m high, spreading to 1.5 m.

Leaves Alternate, reddish with entire margins, soon fall leaving stems bare.

Flowers Reduced, green/yellow, terminal with prominent yellow stamens.

Fruit/seed Round, green, with six longitudinal red stripes. Seeds blackish with a white stripe at the tip.

Flowering period
Most of the year.

Habitat
Dominant shrub on steep rocky mountains and gravel plains; altitude up to 1,500 m.

Distribution
Southern Iran, Oman and Yemen.

Uses
Latex is poisonous and used to kill and catch fish, also as a glue to catch birds. In small quantities is a treatment for parasites in camels.

Family Euphorbiaceae (Spurge Family)

Name

Euphorbia smithii S.Carter

Description

Small deciduous tree, bark peeling on older branches, all parts produce copious white latex when cut; up to 5 m.

Leaves Alternate, oval with a rounded tip and entire margins.

Flowers Terminal clusters with yellow stamens, covered in white hairs.

Fruit/seed A three-lobed hairy capsule with smooth brown or grey seeds.

Flowering period

September–November.

Habitat

Common on the wet slopes of the escarpment in Dhofar; altitude from 50–900 m.

Distribution

Endemic to Dhofar and SE Yemen.

Uses

Latex used to kill skin parasites in goats and camels.

Family Moraceae (Mulberry or Fig Family)

Name

Dorstenia foetida (Forssk.) Schweinf.

Description

A low succulent herb with swollen stems covered in leaf scars; up to 25 cm.

Leaves Elliptic with wavy margins and an acute tip, crowded at the top of the stem.

Flowers Small, sunk into a disc-like receptacle surrounded by leafy rays.

Fruit/seed Small, triangular achene. Seeds ejected explosively.

Flowering period

June–November.

Habitat

In rocky crevices on the escarpment and the drier north facing slopes in Dhofar; altitude from 50–1,100 m.

Distribution

Widespread in the Old World tropics.

Uses

Edible, raw or cooked.

Note

Named after the German botanist Theodore Dorsten (1492–1552).

Family Orchidaceae (Orchid Family)

Name

Epipactis veratrifolia Boiss. & Hohen. ex Boiss.

Description

Erect perennial with leafy stems; up to 60 cm.

Leaves Alternate, lance-shaped with entire margins.

Flowers Green with dark red edges, the lower lip has a white tip; on one side of short stalks arising from leaf nodes.

Fruit/seed Cylindrical capsule, splitting lengthwise. Seeds brown, tiny.

Flowering period

After rain.

Habitat

Damp shady areas, in wadis, near springs and irrigation channels; altitude from 450–2,000 m.

Distribution

NE Africa, Saudi Arabia, Yemen, Turkey, Cyprus, W Asia and east to the Himalayas.

Note

This plant is classified as vulnerable.

Family Orchidaceae (Orchid Family)

Name

Eulophia petersii Rchb.f.

Description
A large orchid with a long flowering spike; up to 3 m.

Leaves Mainly in a basal rosette, lance-shaped, up to 30 cm long with minutely serrated margins.

Flowers Green with purple/brown markings, petals strongly reflexed, at intervals along the erect spike.

Fruit/seed Purple/green, ellipsoid, with tiny dust-like seeds.

Flowering period
September–November.

Habitat
On dry rocky slopes at the base of the escarpment and at the top of north draining wadis in Dhofar; altitude from 200–1,300 m.

Distribution
Tropical and southern Africa, Yemen and Saudi Arabia.

Uses
Juice from the bulbs is used to cure skin infections.

Note
Drought tolerant. *Eulophia* is from the Greek meaning 'beautiful crest'.

Family Piperaceae (Pepper Family)

Name

Peperomia pellucida (L.) Kunth

Description

Small, erect, delicate, epiphytic annual with translucent stems; up to 10 cm.

Leaves Alternate, pale green, translucent, oval with rounded tips and entire margins.

Flowers Minute, on spikes, either axillary or terminal.

Fruit/seed Ovoid, embedded in spikes.

Flowering period

August–October.

Habitat

Damp, shady locations, mainly in the Dhofar wet woodlands, on dead wood or rocks; altitude from 200–900 m.

Distribution

Native to tropical America, now pan-tropical.

Note

Some antibacterial activity has been recorded in the USA.

Family Salvadoraceae (Toothbrush tree Family)

Name

Salvadora persica L.

Common names
Toothbrush tree, arak, rak

Description
An evergreen sprawling shrub with tangled branches that may be erect or pendulous, has an unpleasant smell; up to 2 m high and 6 m wide.

Leaves Opposite, bright green, lance shaped, leathery with entire margins.

Flowers Small, green/white, in terminal or axillary sprays.

Fruit/seed Conspicuous berry-like, turning from white to red or dark purple as they mature.

Flowering period
All year.

Habitat
Common in wadis, sand and gravel areas, often near water, will tolerate some salinity; altitude up to 650 m.

Distribution
Widely distributed from South Africa via the Middle East to India and China.

Uses
Several medicinal uses are reported, including treatment for parasitic infections. Twigs used as toothbrushes.

Note
Attracts the blue spotted arab butterfly (*Colotis phisadia*).

Family Solanaceae (Potato Family)

Name

Withania somnifera (L.) Dunal

Common name
babu

Description
A woody perennial with erect branching stems, covered in white hair; up to 1.5 m.

Leaves Alternate, elliptic with entire margins, densely hairy beneath.

Flowers Small, green/yellow, in clusters of 3–6 in the leaf axils.

Fruit/seed Round berry enclosed in a papery calyx, ripening from green to bright red. Seeds yellow.

Flowering period
Most of the year.

Habitat
In wadis and foothills, often around habitations; altitude up to 1,800 m.

Distribution
Widespread from the Canary Islands to S Asia and Australia.

Note
The name *somnifera* refers to its sleep inducing qualities. Formerly used as a painkiller. *W. qaraitica* is similar.

Family Urticaceae (Nettle or Ramie fibre Family)

Name

Forsskaolea tenacissima L.

Common names
lazzaq, melazaq, tubbaq

Description
A woody based annual or perennial with red stems branching from the base, covered in white hair; up to 60 m.

Leaves Alternate, slightly concave with serrated margins and dense grey hair on the under surface.

Flowers Small, pale green/yellow, clustered at the leaf nodes, enclosed by hairy bracts.

Fruit/seed Brownish achene, enclosed in hairy bracts.

Flowering period
January–May.

Habitat
Common in mountains, wadis, gravel plains and urban areas; altitude up to 1,800 m.

Distribution
SW Europe, N and NE Africa, Arabia, SW Asia.

Note
Named after Forsskål, who died of complications arising from malaria during the Danish Arabia expedition, 1761–1767.

Family Zygophyllaceae (Creosote Bush or Lignum-vitae Family)

Name

Seetzenia lanata (Willd.) Bullock

Description

A prostrate woody-based annual or perennial; up to 30 cm.

Leaves Opposite, trifoliate with pointed tips, on stalks.

Flowers Yellowish green on long stalks from the leaf axils.

Fruit/seed Yellow capsule with five prominent green sections.

Flowering period

February–April.

Habitat

Common in sand and gravel in northern Oman; altitude up to 1,000 m.

Distribution

E and N Africa, Middle East, Arabia, Pakistan and SW India.

Trees

Family Asteraceae (Daisy or Sunflower Family)

Name

Blepharispermum hirtum Oliver

Description
A small, shrubby tree, with greyish bark; up to 3 m.

Leaves Soft, oval with round tips and entire margins.

Flowers A head of small white flowers.

Fruit/seed Black, elliptic achene with hairy edges.

Flowering period
August–October.

Habitat
Common in the wet escarpment woodlands of Dhofar; altitude up to 850 m.

Distribution
Endemic to Dhofar and SE Yemen.

Uses
Firewood, lights easily and is almost smokeless.

Note
Vulnerable, due to its restricted distribution.

Family Avicenniaceae (Black Mangrove Family)

Name

Avicennia marina (Forssk.) Vierh.

Common name
Mangrove

Description
Evergreen with gnarled trunk, vertical air roots appear around the base of trees to absorb oxygen; up to 8 m.

Leaves Opposite, oval with entire margins, dark green on top, pale grey underneath due to excreted salt and fine white hair.

Flowers Small orange/yellow clusters, borne either in the leaf axils or terminally.

Fruit/seed Green, almond-shaped capsules with a single seed.

Flowering period
May–June.

Habitat
Coastal lagoons.

Distribution
Pan-tropical.

Note
Mangroves are very important in marine ecology, giving protection to migrating birds and small marine life. In Oman they are vulnerable, occurring only in small areas of the coast.

Family Bignoniaceae (Trumpet Vine, Jacaranda or Catalpa Family)

Name

Tecomella undulata (Roxb.) Seem.

Common name
ferfer

Description
A tall tree with rough bark; up to 7 m.

Leaves Opposite, oblong with wavy margins.

Flowers Large, orange, trumpet shaped and often grouped together.

Fruit/seed Oblong, slightly curved capsule with winged seeds.

Flowering period
January–May.

Habitat
Wadi beds and near habitation, may be an escaped ornamental; altitude up to 800 m.

Distribution
Arabia to West India.

Family Bombacaceae (Baobab, Balsa, Kapok or Silk-Cotton Family)

Name

Adansonia digitata L.

Common name
Baobab

Description
Large, deciduous tree with a conical grey/black trunk; up to 15 m.

Leaves Alternate, palmate with five leaflets and entire margins.

Flowers Large, white, pendulous, solitary or paired, borne in the leaf axils. Nocturnal, with an unpleasant smell, pollinated by bats.

Fruit/seed Brown, ovoid covered in soft brown hair. Seeds dark brown/black embedded in a white pulp that is edible.

Flowering period
June–August.

Habitat
Only a few trees found in Dalkut and Wadi Hinna in Dhofar province; altitude up to 500 m.

Distribution
Widely distributed in tropical Africa where it grows up to 30 m.

Uses
In Africa many food and other uses are reported.

Note
Named after Michel Adanson, an eighteenth-century French botanist.

Family Burseraceae (Frankincense and Myrrh Family)

Name

Boswellia sacra Flueck.

Common names
Frankincense, luban

Description
Small tree with papery peeling bark, all parts contain resin; up to 5 m.

Leaves Alternate, compound, crowded at the end of the branches. Paired leaflets have wavy margins and increase in size towards the tip, the terminal leaflet being the largest.

Flowers White, borne on axillary spikes at the end of the branches with a sweet honey scent.

Fruit/seed Oval capsule, green turning red/brown when mature.

Flowering period
March–May.

Habitat
The arid mountainous zone in Dhofar beyond the monsoon moisture but within the range of the cooling winds; altitude up to 1,000 m.

Distribution
Tropical NE Africa to southern Arabia.

Note
Source of Frankincense. Wild trees are over-exploited making it vulnerable.

Family Burseraceae (Frankincense and Myrrh Family)

Name

Commiphora gileadensis (L.) C.Chr.

Description

Large shrub or tree with peeling bark and fragrant resin when cut; up to 4 m.

Leaves Alternate, compound having 3–5 oblong leaflets with entire margins, hairy.

Flowers Bright red with four petals, small, borne on side shoots among the leaves.

Fruit/seed Dull red, ovoid with four longitudinal white stripes.

Flowering period

Most of the year.

Habitat

Common in Dhofar, ranging from a small shrub in dry coastal areas to a tall tree on the escarpment; altitude up to 850 m.

Distribution

SW Arabia, NE and E Africa.

Uses

Fruit is edible and said to be a thirst quencher. The heartwood was used as a yellow dye. Dried bark reported to be a disinfectant for large wounds.

Note

Several species of *Commiphora* are found in Oman. Attracts the yellow patch white butterfly (*Colotis halimede*).

Family Capparaceae (Caper Family)

Name

Boscia arabica Pestal.

Description
Small evergreen, with a dense flat crown, bark smooth, greyish/brown; up to 5 m.

Leaves Spirally arranged on short compact branches, oval with entire margins.

Flowers Yellow-green, solitary or in clusters, borne in the leaf axils on erect stems.

Fruit/seed Ovoid, grape-like fruit turning from green to pale yellow as they mature.

Flowering period
September.

Habitat
Characteristic of the escarpment woodlands in Dhofar; altitude from 80–700 m.

Distribution
Regional endemic to southern Arabia; from Dhofar to Aden.

Uses
The wood is termite resistant and therefore useful for building.

Note
This tree is becoming scarce and may be threatened. Attracts several butterflies including the caper white (*Anaphaeis aurota*), zebra white (*Pinacopteryx eripha*) and yellow patch white (*Colotis halimede*).

Family Capparaceae (Caper Family)

Name

Maerua crassifolia Forssk.

Common name
sarh

Description
Evergreen, mature trees have a rounded or flattened crown, young trees appear as bushy shrubs with drooping branches, bark grey; up to 7 m.

Leaves Spirally arranged in clusters, small, oval, somewhat fleshy.
Flowers In small clusters, without petals, numerous long white stamens radiating from backward curving green sepals.
Fruit/seed Long cylindrical pods constricted between the seeds, green, turning brown when ripe.

Flowering period
Most of the year.

Habitat
Mountain slopes and gravel plains throughout Oman; altitude up to 1,500 m.

Distribution
Tropical Africa, Arabia and east to Pakistan.

Uses
In Africa boiled leaves are said to ease constipation.

Note
Attracts caper white (*Anaphaeis aurota)* and golden arab (*Colotis chrysonome*) butterflies.

Family Celastraceae (Spindle tree Family)

Name

Maytenus dhofarensis Sebsebe

Description
Small tree or shrub, with spiny stems, often forming a thicket; up to 3 m.

Leaves Alternate, dark green, elliptical with entire, or very slightly serrated margins.

Flowers Small, white flowers, either solitary or clustered.

Fruit/seed Red, triangular capsule with brown seeds.

Flowering period
Most of the year.

Habitat
Throughout the escarpment woodlands, extending into the drier plateau; altitude up to 1,250 m.

Distribution
Endemic to Dhofar and SE Yemen.

Uses
Thorns are vicious, which makes this a good fencing plant.

Note
Believed to be a haunt of snakes.
M. senegalensis is similar and found throughout Africa, the Middle East and east to India.

Family Combretaceae

Name

Anogeissus dhofarica A.J.Scott

Common names
misteh, meset

Description
Deciduous tree with a spreading crown and rough, dark brown bark; up to 10 m.

Leaves Blue/green, opposite, elliptical with entire margins, young leaves are hairy.

Flowers Yellow clusters borne at the end of short axillary branches.

Fruit/seed Cone-shaped ball, pale brown, containing many winged seeds.

Flowering period
July–August.

Habitat
A dominant species on the Dhofar escarpment; altitude up to 1,200 m.

Distribution
Endemic to Dhofar and SE Yemen. *Anogeissus* species are found from E Africa to Thailand.

Uses
Widely used for building and firewood. An important source of fodder in the dry season.

Note
Due to overuse this tree is becoming vulnerable.

Family	Cupressaceae (Conifer Family)

Name

Juniperus excelsa M.Bieb.

Common names
Juniper, al alan

Description
A gnarled, evergreen tree; up to 10 m.

Leaves Young leaves needle-like, adult leaves scaly.

Flowers Absent, male and female cones.

Fruit/seed Blue/black, berry-like with a whitish waxy bloom.

Habitat
Dominant tree on Jabal Al Akhdar; altitude above 1,300 m.

Distribution
Eastern Mediterranean to Pakistan, where it may grow up to 20 m.

Note
Often found growing with *Sideroxylon mascatense* and *Olea europaea*. May be vulnerable in Oman due to being restricted to a small area.

Family Dracaenaceae (Dragon-Tree Family)

Name

Dracaena serrulata Baker

Common name
ariyeb

Description
Single trunk, forked above; up to 5 m.

Leaves Rigid, long and thin, densely congested at the end of branches.

Flowers Tiny, bunched at the end of branches.

Fruit/seed Round, fleshy.

Flowering period
June–August.

Habitat
The dry plateau at the top of the escarpment and north facing hills in Dhofar, beyond the range of the monsoon; altitude from 800–1,400 m.

Distribution
Endemic to dry areas of SW Arabia; closely related species are found on the African mainland.

Uses
Fibre from the leaves used to make strong ropes.

Note
This species is endangered and should be protected.

Family Leguminosae (Pea, Bean or Mimosa Family)

Name

Acacia spp.

Description

A large genus of over 1,450 species of mainly evergreen trees or shrubs, conspicuous spines may be straight or curved; up to 20 m.

Leaves Normally bi-pinnate with up to 13 pairs of leaflets, oval with smooth margins.

Flowers Either yellow or white, on spikes or as small balls along the branches.

Fruit Round or flattened pods, which may be twisted.

Flowering period

After rain.

Habitat

Open savannah areas; altitude up to 2,000 m.

Distribution

Africa, Arabia and Asia. Common in Australia, where they are known as Wattles. There are 13 species of *Acacia* in Oman.

Name

Acacia tortilis (Forssk.) Hayne

Common name

samur

Small branched tree with a dense flattened crown; up to 6 m. Flowers white balls with a sweet scent. Fruit twisted green pods, turning brown when mature. An important forage tree; the pods contain up to 18 percent protein.

Family Leguminosae (Pea, Bean or Mimosa Family)

Name

Acacia ehrenbergiana Hayne

Common name

salam

A shrub that forms a mound when growing in sand. Leaves are paired, each with 8–12 opposite leaflets. Flowers are yellow balls.

Family Leguminosae (Pea, Bean or Mimosa Family)

Name

Acacia gerrardii Benth.

Common name

tulh

Medium sized tree: up to 5 m. Bark blackish. Flowers white balls.

Family Leguminosae (Pea, Bean or Mimosa Family)

Name

Acacia nilotica (L.) Delile

Common names

qarat, tulh

Tree with grey bark and prominent straight white spines. Flowers yellow balls. Fruit constricted between seeds.

Name

Acacia senegal (L.) Willd.

Common name

thor

A tree with white flowers borne on a spike. Three prickles in the leaf nodes; the central one turned downwards and the other two upwards. Fruit is flattened and relatively broad. Resin is a commercial source of gum Arabic.

Family Leguminosae (Pea, Bean or Mimosa Family)

Name

Delonix elata (L.) Gamble

Description
Deciduous tree with a round spreading crown; up to 8 m.

Leaves Alternate, compound, bipinnate with up to 12 pairs of leaflets.

Flowers Large, showy white, turning yellow after pollination, with prominent stamens.

Fruit/seed Flattened oblong pod with an acute tip. Seeds pale brown.

Flowering period
September–November.

Habitat
Common in the monsoon area of the Dhofar escarpment; altitude up to 850 m.

Distribution
Widely distributed in dry hilly country from tropical Africa to India.

Uses
Used for the production of lime. Pods eaten raw.

Family Leguminosae (Pea, Bean or Mimosa Family)

Name

Prosopis cineraria (L.) Druce

Common name
ghaf

Description
Large tree, with a dense, often rounded crown, branches droop and bear many short spines between the leaf nodes, grey bark is fissured; up to 10 m.

Leaves Alternate, grey-green, compound, divided into two pinnae each with 6–16 paired leaflets.

Flowers Small, pale yellow, on cylindrical spikes borne in leaf axils, pendent.

Fruit/seed Reddish brown pods, slightly curved, containing many seeds.

Flowering period
January–May.

Habitat
Dominant tree of sand deserts, also found along wadi channels and on gravel plains; altitude up to 150 m.

Distribution
Arabia and east to India.

Uses
Important in desert areas as fodder for camels, the wood is used for fuel and construction. A variety of medicinal uses are reported, mainly as a pain killer.

Family Leguminosae (Pea, Bean or Mimosa Family)

Name
Prosopis juliflora (Sw.) DC.

Common name
ghaf al bahri

Description
Small spreading evergreen tree, may be branched from close to the ground, bark smooth, light brown with spines; up to 10 m.

Leaves Alternate, bright green, compound with up to 22 pairs of opposite leaflets.

Flowers Yellow to white, small, on cylindrical hanging spikes up to 10 cm long.

Fruit/seed Long, bright green pods, sometimes slightly twisted, hang in clusters.

Flowering period
Most of the year.

Habitat
Sand and gravel areas, often on waste ground; altitude up to 250 m.

Distribution
Native to Mexico, now widely distributed in arid and semi-arid regions of the tropics.

Note
Drought resistant and tolerant of salinity. Invasive, originally introduced as an ornamental, now competing with *P. cineraria* and other native tree species. Should not be propagated.

Family	Leguminosae (Pea, Bean or Mimosa Family)

Name

Tamarindus indica L.

Common name
Tamarind

Description Evergreen, rough greyish bark; up to 12 m.

Leaves Alternate, bright green, compound with 8–20 paired leaflets.

Flowers Large, cream with red veins on stalks.

Fruit/seed Large brown pods. Seeds embedded in chestnut-brown pulp, sour to taste.

Flowering period
April–June.

Habitat
Often near water, on sea-facing slopes on the monsoon affected escarpment in Dhofar; altitude up to 800 m.

Distribution
Native to tropical Africa, naturalised throughout the tropics.

Uses
Fruit is edible, high in protein and vitamin C, acts as a laxative if eaten in excess, useful fodder in the spring.

Note
Tolerant of a wide range of conditions. A common ornamental and fruit tree in tropical countries.

Family Moraceae (Mulberry or Fig Family)

Name

Ficus cordata Thunb.

Common name
Wild fig, lithab

Description
Tree, branching from the base, with smooth, grey bark; up to 12 m.

Leaves Alternate on long pale stalks, lance-shaped with entire margins.

Flowers Minute, grow within the fruit.

Fruit/seed Urn shaped or round, resembling the common fig, but smaller.

Flowering period
Most of the year.

Habitat
Near water in wadis and urban areas; altitude up to 2,000 m.

Distribution
Africa, Saudi Arabia, UAE and Yemen.

Uses
Some medicinal uses are reported, for example the treatment of scorpion stings in the UAE.

Note
Fruit that fall into water are eaten by fish. There are seven species of wild *Ficus* in Oman, in addition there are several cultivated species, often grown for shade.

Family Moringaceae (Horseradish tree, Drumstick tree or Ben-oil tree Family)

Name

Moringa peregrina (Forssk.) Fiori

Common names
Wild drumstick tree, shua

Description
Deciduous tree with green drooping branches; up to 6 m.

Leaves Oblong, drop as soon as the flowers appear leaving the branches leafless.

Flowers Fragrant, white/pink, occur in sprays from the leaf axils.

Fruit/seed Long, cylindrical pods with prominent longitudinal ridges, hang down. Seeds large and brown.

Flowering period
February–May.

Habitat
Common on mountain slopes throughout Oman; altitude up to 1,200 m.

Distribution
NE Africa, Arabia and SW Asia.

Uses
Oil from the seed has a variety of medicinal uses.

Note
The wood is termite resistant.

Family Oleaceae (Olive Family)

Name

Olea europaea L.

Common names
Olive, itm

Description
Small tree; up to 7 m.

Leaves Opposite, lance-shaped, shiny grey/green above and brownish green below with entire margins.

Flowers Creamy white, fragrant, borne in the leaf axils.

Fruit/seed Purple/black when ripe with a single seed.

Flowering period
May–August.

Habitat
Common in the northern mountains and near the top of the escarpment in Dhofar, often found growing with *Juniperus excelsa* and *Sideroxylon mascatense*; altitude above 600 m.

Distribution
Widespread from the Mediterranean to Australasia.

Uses
Has many culinary and medicinal uses. Fodder for goats.

Family Rhamnaceae (Buckthorn or Jujube Family)

Name

Ziziphus spina-christi (L.) Desf.

Common names
Christ's thorn, sidr (plant), nabaq (fruit)

Description
Tree with a spreading crown, trunk may be divided near the base; up to 10 m.

Leaves Alternate, light green, oval with entire margins and three conspicuous veins rising from the base. Paired spines at the base of the leaves, one straight and the other hooked (may be absent in cultivated plants).

Flowers Cream with five petals around a central disc, in dense clusters in the leaf axils, fragrant.

Fruit/seed Round green/yellow, turning reddish brown when mature.

Flowering period
February–April, or longer when in cultivation.

Habitat
Widely distributed in wadis and gravel plains, common as a street tree in Muscat; altitude up to 1,400 m.

Distribution
Naturalised from NE Africa to NW India.

Note

Mentioned in the Qur'an in a description of paradise. The epithet *spina-christi* refers to the crown of thorns worn by Christ at the crucifixion. Attracts the Mediterranean pierrot butterfly (*Tarucus rosaceus*).

Family Sapotaceae (Sapodilla, Chicle, Gutta-Percha, or Mastic Family)

Name

Sideroxylon mascatense (A.DC.) T.D.Penn.

Common name

but

Description

Small spiny tree or large shrub, with rather grey, gnarled branches ending in spines; up to 2 m.

Leaves Alternate, grey/green, oval with entire margins and a pointed tip.

Flowers Small, white or pale yellow, with protruding stamens, clustered at the end of branches.

Fruit/seed Berry turning purple or black when ripe, edible.

Flowering period

April–June.

Habitat

The northern mountains, often growing in close association with *Olea europea* and *Juniperus excelsa*; altitude from 1,000–2,500 m.

Distribution

NE Africa, Arabia, Iran, Afghanistan and SW Pakistan.

Grasses

Family Poaceae

Name

Arundinella pumila (Hochst. ex A.Rich.) Steud.

Habitat

Shady escarpment woodlands and tall grassland in Dhofar; altitude from 300–900 m.

Distribution

Tropical Africa, Oman, tropical Asia.

Family Poaceae

Name

Cenchrus ciliaris L.

Habitat

Sand and gravel wadis and mountains, along roadsides and as a weed of cultivation; altitude from 0–2,400 m.

Distribution

Tropical Africa, Arabia, east to India; widely introduced in the Old World.

Family Poaceae

Name

Chloris barbata Sw.

Habitat

Sand and gravel wadis and mountains, along roadsides and as a weed of cultivation; altitude from 0–1,050 m.

Distribution

Throughout the tropics.

Family Poaceae

Name

Cymbopogon schoenanthus

(L.) Spreng.

Habitat

Sandy and gravelly plains and wadi beds, and rocky hillsides; altitude from 0–2,200 m.

Distribution

N and E Africa, and Arabia.

Note

May be difficult to distinguish from the closely related *Cymbopogon commutatus*.

Family Poaceae

Name

Dactyloctenium aegyptium (L.) P.Beauv.

Habitat

Sand and gravel areas, along roadsides and as a weed of cultivation; altitude from 0–1,500 m.

Distribution

Old World tropics.

Family Poaceae

Name

Lasiurus scindicus Henrard

Habitat

Sand and gravel wadis and mountain slopes; altitude from 0–1,050 m.

Distribution

Africa, Arabia, east to India.

Family Poaceae

Name

Pennisetum setaceum
(Forssk.) Chiov.

Habitat
Gravel wadis and mountain slopes, in crevices and ravines; altitude from 0–2,500 m.

Distribution
N and NE Africa, Arabia, SE Asia.

Family Poaceae

Name

Saccharum ravennae (L.) Murr.

Habitat

In wadis and near water; altitude 50-1,000 m.

Distribution

Mediterranean, N Africa, Arabia, eastwards to India.

Family Poaceae

Name

Setaria verticillata (L.) P. Beauv.

Habitat

Damp and shady locations, in gardens and plantations; altitude 0–2,200 m.

Distribution

Tropical and warm temperate regions throughout the world.

Family Poaceae

Name

Stipagrostis uniplumis
(Licht.) de Winter

Habitat
In sandy, stony and rocky wadis; altitude 0–1,900 m.

Distribution
Tropical and southern Africa to India.

Family Typhaceae

Name

Typha domingensis Pers.

Habitat

In wadis and near water; altitude 0–2,000 m.

Distribution

Pantropical.

Family Cyperaceae

Name

Cyperus conglomeratus Rottb.

Habitat

Coastal sandy areas, sandy and gravelly soils; altitude 0–300 m.

Distribution

Dry regions of Africa, Arabia and SE Asia.

Family Juncaceae

Name

Juncus rigidus Desf.

Habitat

Common near streams and permanent water sources; altitude 0–200 m.

Distribution

Mediterranean, N Africa, Arabia and eastwards to Iran, Afghanistan and Pakistan.

Family Adiantaceae

Name

Adiantum capillus-veneris L.

Habitat

Damp shady places; altitude 0–3,000 m.

Distribution

Widely distributed in warm temperate and tropical areas of the world.

Family Pteridaceae

Name

Pteris vittata L.

Habitat

Damp shady places, terrace walls, wadi sides and cliffs; altitude 0–2,500 m.

Distribution

Southern and tropical Africa, Arabia.

Family Selaginellaceae

Name

Selaginella imbricata (Forssk.)Spring ex Decne.

Habitat

Damp shady areas in Dhofar; altitude from 100–1,500 m.

Distribution

Southern and eastern Africa, Arabia.

Glossary

Achene a small dry, one-seeded, closed fruit.

Alternate leaves arranged alternately on the stem.

Annual a plant that completes its life cycle in one year, or one growing season.

Appendage an attached subsidiary part.

Ascending directed upwards at an oblique angle.

Axil/axillary the angle between the stem and leaf from which buds of new leaves or inflorescences grow.

Basal originating at the base.

Beak a pointed projection.

Berry a fleshy, indehiscent fruit with the seed(s) immersed in pulp.

Bi-lobed with two lobes.

Blade the expanded part of the leaf usually attached to the stem by a short stalk.

Bract a modified leaf attached below a flower, often brightly coloured.

Calyx the outer whorl of floral parts, usually green, that protect the flower.

Capsule a dry, splitting fruit with two or more seed-containing compartments.

Compound leaf a leaf divided into a number of smaller leaflets.

Conifer trees or shrubs bearing cones, mostly with needle or scale-like leaves.

Corona a crown like outgrowth from a corolla tube.

Corolla a collective term for petals which may be free or fused into a corolla tube.

Cosmopolitan worldwide distribution.

Crest an elevation or ridge.

Deciduous shedding leaves and/or other organs annually.

Decumbent running along the ground, with the end ascending.

Dehiscent splitting open.

Double flowered a corolla with two or more times the normal number of petals.

Elliptic a leaf shape that is much longer than it is wide.

Ellipsoid elliptical with a three dimensional body.

Endemic restricted to a specific area – not occurring elsewhere.

Entire without any breaks or serrations to leaf margins.

Epicalyx an extra whorl of appendages below the calyx.

Epiphyte a plant growing on another plant but not using the host plant for food.

Follicle a dry fruit with only one seed compartment splitting along the suture.

Gland a small superficial organ that secretes oils and other substances.

Herbaceous mainly small annual or perennial plants; not woody.

Indehiscent remaining closed at maturity (fruit).

Inflorescence a cluster of flowers.

Latex a fluid produced by certain plants, which contains various substances such as alkaloids, minerals etc.

Leaflet a leaf-like segment of a compound leaf.

Lip/labellum one of three petals in an orchid that differs from the other two and gives the flower its characteristic form. Always distinct and often larger than the other petals and sepals.

Lobe(d) a division of an organ, generally rounded.

Mid-vein the main central vein of the leaf that is a continuation of the veins in the petiole.

Mucilaginous slimy.

Native belonging to a region through natural circumstances (not introduced).

Naturalized introduced into a region and subsequently surviving naturally.

Node point of attachment of leaves to the stem.

Nutlet a small nut.

Obovate oval but with the narrow end attached to the stalk.

Opposite leaves arranged as pairs on opposite sides of the stem, arising from the same node.

Orbicular round.

Ovoid egg shaped.

Palmate (leaves) a compound leaf having four or more leaflets arising from a single point. (**Veins**) several equally prominent veins branching out from the base of a leaf blade.

Parallel venation leaf veins lie parallel to each other.

Parasitic deriving nutrients from another plant.

Pendent hanging down.

Perennial a plant with an indefinite life span, taking more than a year to grow and which flowers for a long period of time.

Petiole the stalk of a leaf.

Pinna the first division of a compound leaf.

Pinnate two rows of leaflets, one along either side of a mid-rib, feather like.

Pod a dry splitting fruit, opening along two sutures.

Prickle a sharp pointed outgrowth from the superficial tissues of the stem.

Prostrate a plant that grows closely along the ground.

Types of leaf

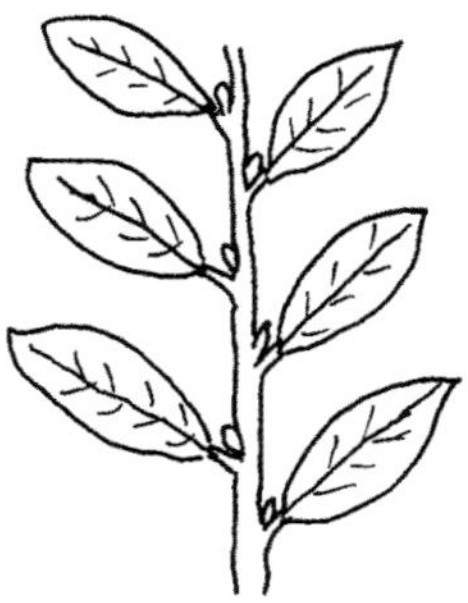

ALTERNATE

COMPOUND

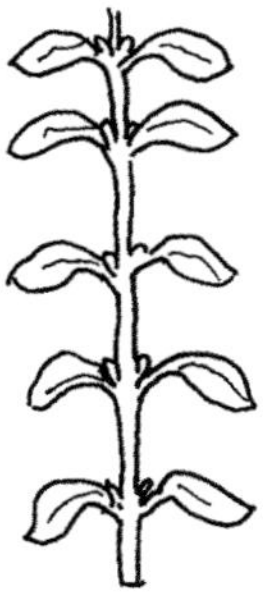

OPPOSITE

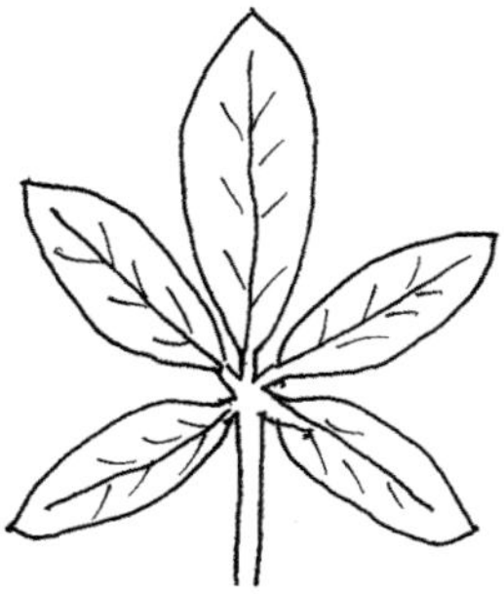

PALMATE

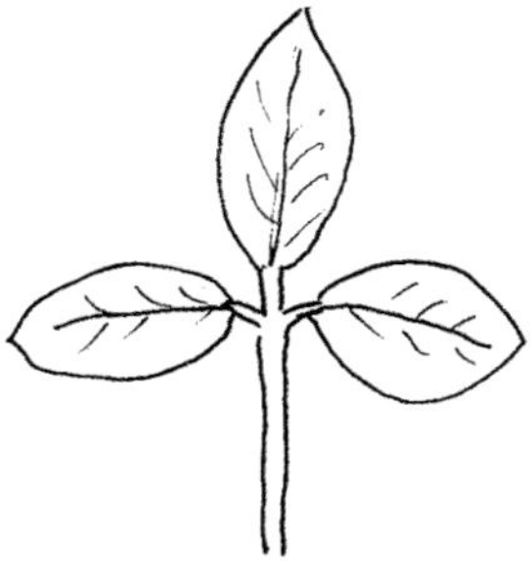

TRIFOLIATE

WHORLED

Reflexed bent backwards.

Resin a sticky, often fragrant substance that is insoluble in water, secreted by specialised cells.

Rhizome an underground stem that grows horizontally.

Rosette a tight spiral of basal leaves.

Sap the juice of a plant.

Scale a reduced leaf.

Sepals the segments of the calyx.

Serrated leaves with a saw-toothed margin.

Shrub a perennial woody plant with several main stems.

Simple leaf a single leaf arising from the axil.

Species Linnaean unit of classification; a group of plants of similar shape and constant distinctive characteristics.

Spike flowers on a central stem without stalks.

Spine a sharp woody outgrowth from a leaf.

Stamens the male parts of a flower consisting of stalk (filament) and anther (pollen sac).

Stellate star shaped.

Stem the main supporting axis of a plant.

Succulent a plant with fleshy leaves.

Tendril a thread like structure that grasps other plants or objects and clings to them for support.

Terminal at the end of a stem or at the extreme top.

Thorn a short pointed branch.

Tree a perennial woody plant, at least 2 m tall, with a main trunk.

Trifoliate with three leaves.

Tubular cylindrical or hollow.

Venation arrangement of veins in a leaf.

Whorled a circular arrangement of leaves with more than two to a node.

Xerophytic a plant which can live in very dry conditions.

Bibliography

Bhandari, M.M. (1995). *Flora of the Indian Desert,* MPS Repros, Jodhpur.

Blundell, M. (1992). *Collins Photo Guide to the Wild Flowers of East Africa*. Harper Collins, London.

Clapp, N. (1999). *The Road to Ubar*. Souvenir Press, London.

Fiennes, R. (1992). *Atlantis of the Sands: The search for the Lost City of Ubar*. Bloomsbury Publishing, London.

Ghazanfar, S.A. (1992). *An Annotated Catalogue of the Vascular Plants of Oman and their Vernacular Names*. National Botanic Garden of Belgium, Meise, Belgium.

Ghazanfar, S.A. (ed). (2003). *Flora of the Sultanate of Oman*. Vol. 1: Piperaceae–Primulaceae (text + photo CD). National Botanic Garden of Belgium, Meise, Belgium.

Ghazanfar, S.A. (ed). (2007). *Flora of the Sultanate of Oman*. Vol. 2: Crassulaceae– Apiaceae (text + photo CD). National Botanic Garden of Belgium, Meise, Belgium.

Ghazanfar, S.A. & Fisher, M. (eds). (1998). *Vegetation of the Arabian Peninsula*. Kluwer Academic Press, The Netherlands.

Heywood, V.H., Brummitt, R.K., Culham, A. & Seberg, O. (2007). *Flowering Plant Families of the World*. Royal Botanic Gardens, Kew.

Jongbloed, M. (2003). *A Comprehensive Guide to the Wildflowers of the United Arab Emirates*. Environmental Research and Wildlife Development Agency, Abu Dhabi.

Larsen, T & K. (1980). *Butterflies of Oman*. John Bartholomew & Son Ltd, London.

Mandaville, J. (1978) *Wild Flowers of Northern Oman*, John Bartholomew & Son Ltd. Bromley, Kent.

Miller, A.G. & Morris, M. (1988). *Plants of Dhofar*. The Office of The Adviser for Conservation of The Environment, Diwan of Royal Court, Sultanate of Oman, 1988.

Miller, A.G. & Cope, T. (1996). *Flora of the Arabian Peninsula and Socotra*. Vol. 1. Edinburgh University Press, Edinburgh.

Patzelt, A. (2007). *Oman Plant Red Data Book*. The Office of The Adviser for Conservation of The Environment, Diwan of Royal Court, Sultanate of Oman.

Thulin, M. (ed). (1993–2006). *Flora of Somalia*. Vols 1–4. Royal Botanic Gardens, Kew.

Wood, J.R.I. (1997). *Handbook of the Flora of Yemen*. Royal Botanic Gardens, Kew.

www.ansci.cornell.edu/plants/toxicagents

www.efloras.org

www.hear.org

www.ildis.org/Leguminosae

www.ipni.org

www.kew.org

www.mobot.org

www.plants.usda.gov

Index of Arabic Names

Index of common and scientific names